Basic Electricity

REVISED EDITION

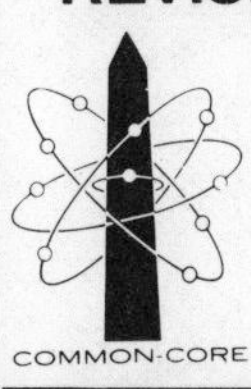

VAN VALKENBURGH,
NOOGER & NEVILLE, INC.

VOL. 4

Hayden Book Company
A DIVISION OF HAYDEN PUBLISHING COMPANY, INC.
HASBROUCK HEIGHTS, NEW JERSEY / BERKELEY, CALIFORNIA

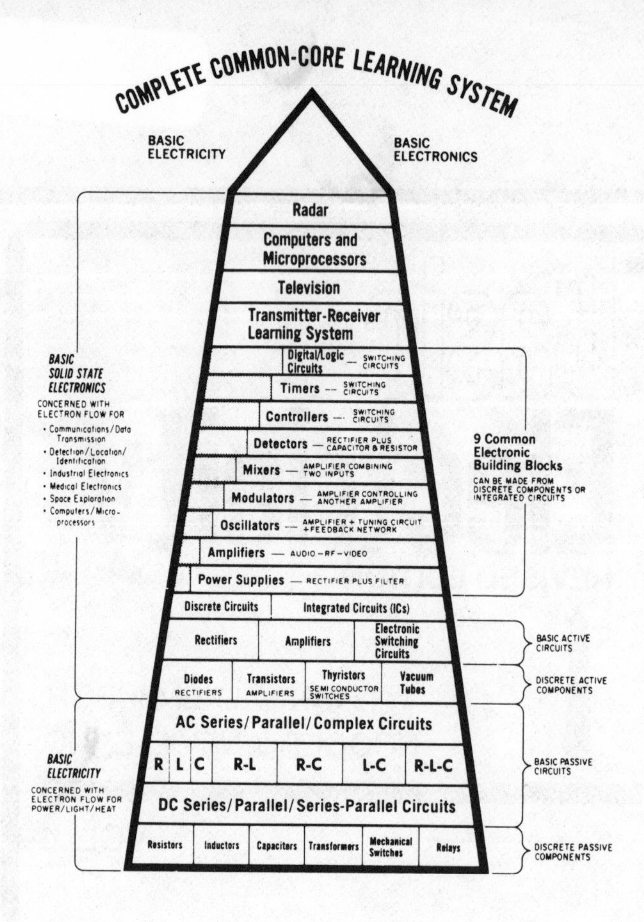

ISBN 0-8104-0879-1
Library of Congress Catalog Card Number 76-57841

Printed in the United States of America

7 8 9 PRINTING

85 86 87 88 89 90 91 YEAR

Preface to Revision of BASIC ELECTRICITY

The COMMON-CORE® Program – *Basic Electricity, Basic Electronics, Basic Synchros and Servomechanisms,* etc. – was designed and developed during the years 1952–1954. On the basis of a job task analysis of a broad spectrum of U.S. Navy electrical/electronics equipment of that era, there was established a "common-core" of prerequisite knowledge and skills. This "common-core" prerequisite was then programed into a teaching/learning system which had as its primary instructional objective the effective training of U.S. Navy electrical/electronics technicians who could understand and apply such understanding in meaningful job problem situations.

Since that time, over 100,000 U.S. Navy technicians have been efficiently trained by this performance-based system. Civilian students and technicians have accounted for hundreds of thousands more. The military and civilian education and training programs in South America, Europe, the Middle East, Asia, Australia, and Africa have also recognized its usefulness with some 12 foreign-language editions presently in print.

Now the foundation of the COMMON-CORE Program, *Basic Electricity,* is being updated and improved. Its equipment job task base has been enlarged to cover the understanding and skills needed for the spectrum of present-day electrical/electronic equipment – modern industrial machines, controls, instrumentation, computers, communications, radar, lasers, etc. Its technological components/circuits/functions base has been revised and broadened to incorporate the generations of development in electrical/electronics technology – namely, from (1) vacuum tubes to (2) transistors and semiconductors to (3) integrated circuits, large scale integration, and microminiaturization.

Educationally, considerable effort has been given to incorporating individualized learning/testing features and techniques within the texts themselves, and in the accompanying interactive student mastery tests.

Notwithstanding the passage of time, the original innovative, basic text-format, system-design elements of the COMMON-CORE Program still stand – this solid framework of proved effectiveness that has been the stimulus for many of the improvements in vocational/technical education.

VAN VALKENBURGH, NOOGER & NEVILLE, INC.

New York, N.Y.

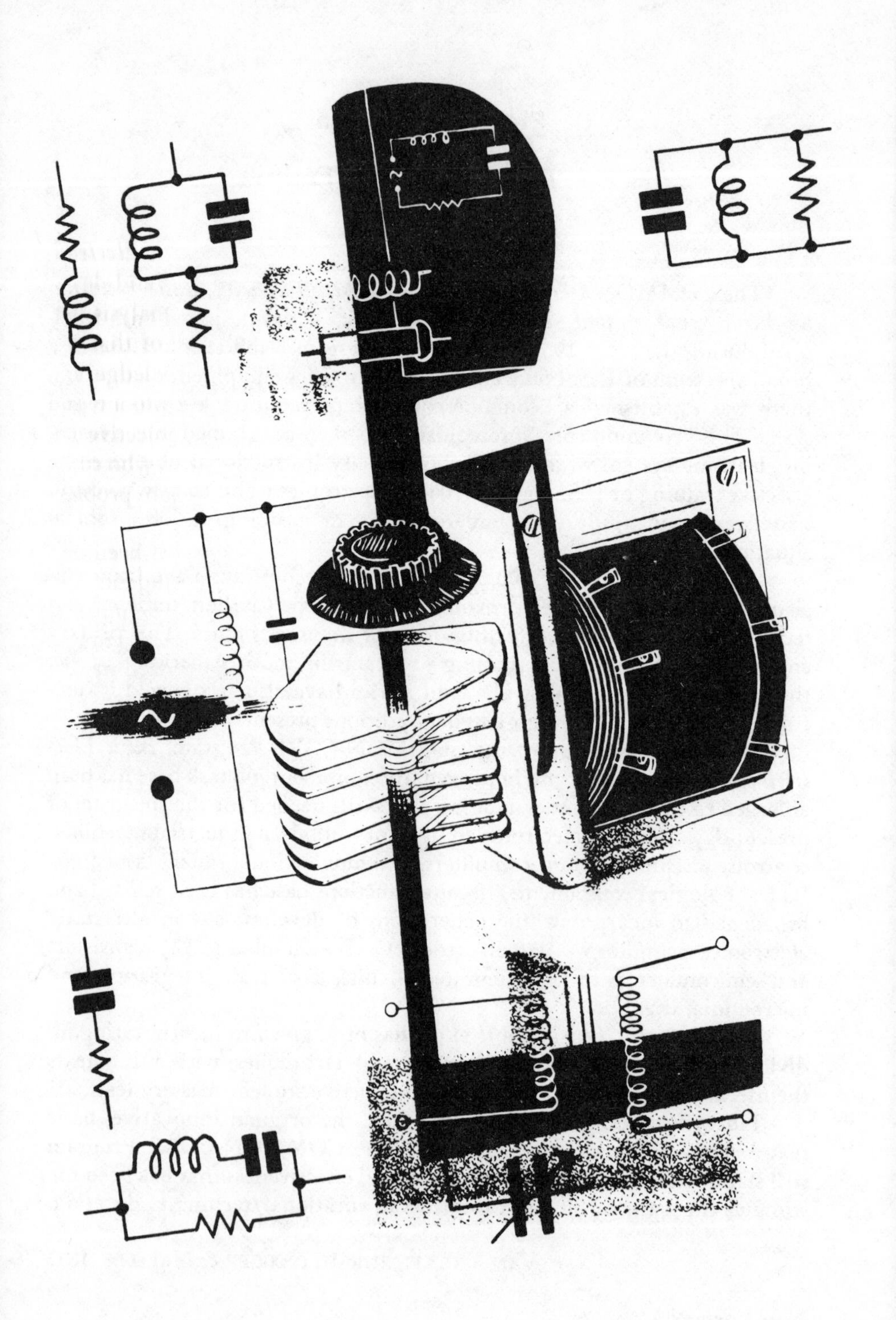

CONTENTS

Transformers

Introduction to AC Power Distribution

Troubleshooting AC Circuits

Appendix

Introduction to AC Electric Circuits

Now that you know about direct current (dc) and alternating current (ac) and how they work from Volumes 2 and 3, and how resistance (R), inductance (L), and capacitance (C) behave in basic circuits carrying dc and ac, you are ready to learn how all three of these circuit elements can be used in combination to control and influence current flow in ac electric circuits. You also have learned how to analyze simple ac (and dc) circuits that contain *only one* of these circuit elements—resistance, inductance, or capacitance alone. You have found out how R, L, and C *individually affect* ac current flow, phase angle, and power in ac circuits. However, you have not as yet considered ac circuits having *two or more* of these circuit elements. So, here, in Volume 4, we now are going to put them *all together*!

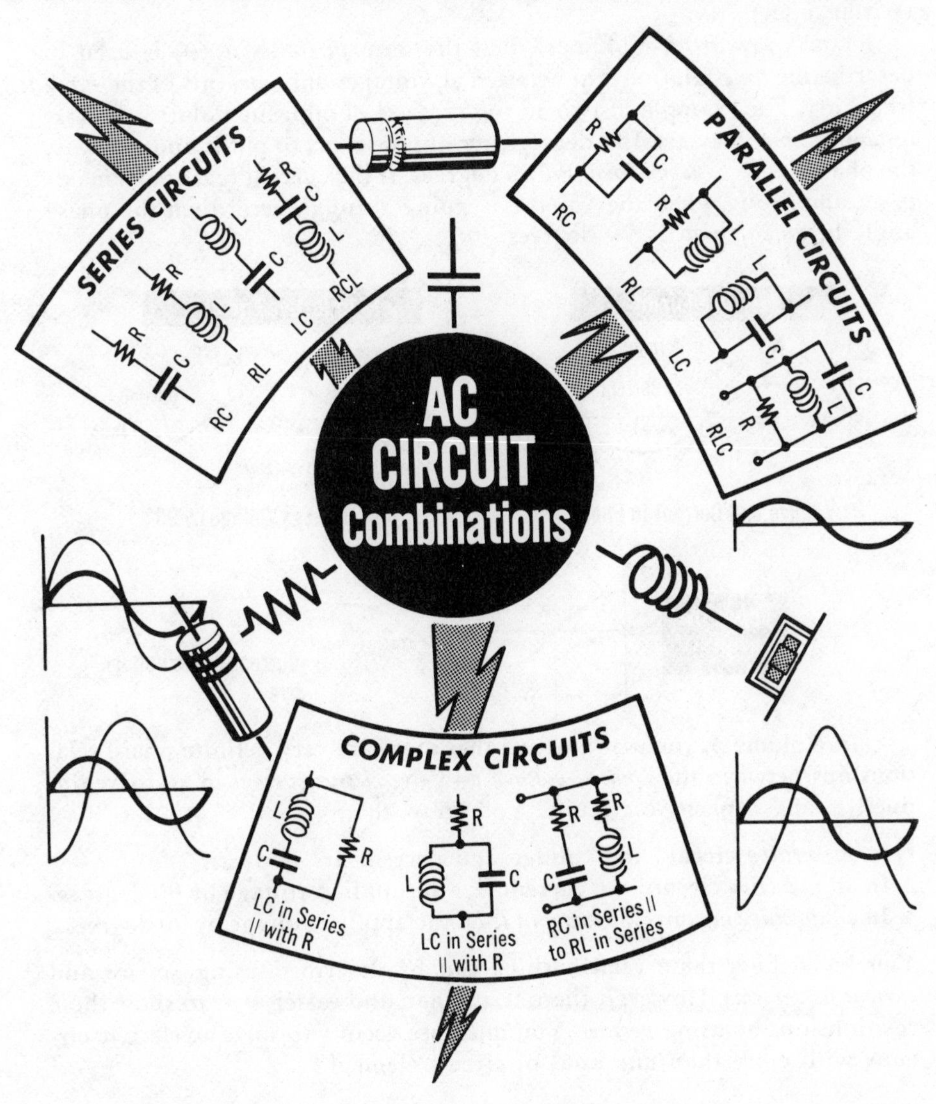

Introduction to AC Electric Circuits (continued)

In this volume you will learn how to analyze, solve, and troubleshoot ac circuits that contain both series and parallel combinations of the three circuit elements—resistance (R), inductance (L), and capacitance (C).

As you know, every electric circuit contains a certain amount of R, L, and C and inductive and capacitive reactance, X_L and X_C. Therefore, ac circuits can contain *three* factors which oppose current flow—R, X_L, and X_C. (In any given circuit, however, if any of these factors is negligible, it can be disregarded.) As you will learn in this volume, these factors must be combined in special ways to find the *net* opposition to the flow of current. This combined factor, called *impedance*, is represented by the symbol *Z*. The Ohm's law formula is used to calculate Z as you used it for finding resistance (R).

You know from Volume 3 that the term *phase angle* (θ) is used to describe the *time* relationship between ac voltages and currents of the same frequency. For example, if two ac voltages are of opposite polarity at *every instant* of *time*, they are 180 degrees out of phase, or, to put it another way, the phase angle *between them* is 180 degrees. If the current reaches its maximum amplitude when the voltage is going through zero, then the phase angle between them is 90 degrees.

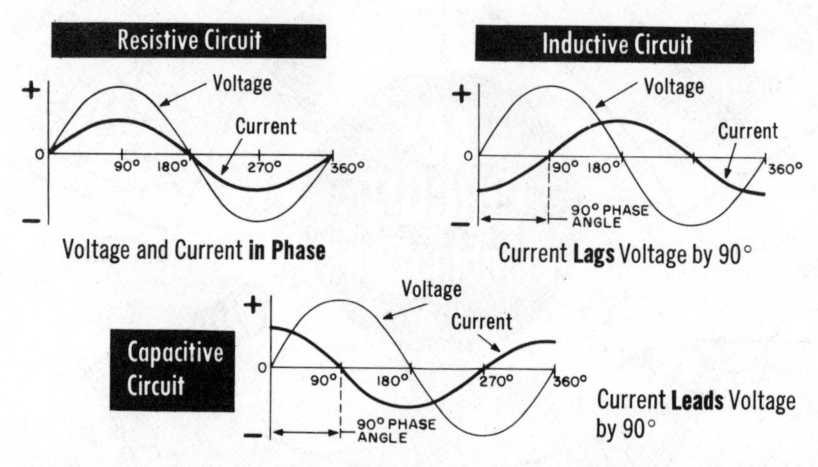

In Volume 3, you also learned that there are very definite phase relationships between the *applied voltage* and the *circuit current* in resistive, inductive, and capacitive circuits. You know that:

1. In a *resistive* circuit, the voltage and current are *in phase*.
2. In an *inductive* circuit, the current *lags* the applied voltage by 90 degrees.
3. In a *capacitive* circuit, the current *leads* the applied voltage by 90 degrees.

You know how these relationships can be described using *voltage* and *current waveforms*. However, there is another, and easier, way to show these relationships by using *vectors*. You must use vectors to solve ac electric circuits with more than one kind of circuit element.

Introduction to AC Electric Circuits (continued)

Since you will be thinking now about the interrelationship of all these various factors, you will need some additional tools to help you keep them all in mind and understand them. The additional tools that you are going to learn about as you continue to study ac electric circuits—starting with ac series circuits—are the following: (1) Use of *vectors* in the solution of ac circuits, and (2) solution of ac circuit problems either *graphically* or by *calculation*. As you will see, solving ac circuits by calculation will be made much easier if you have a hand calculator.

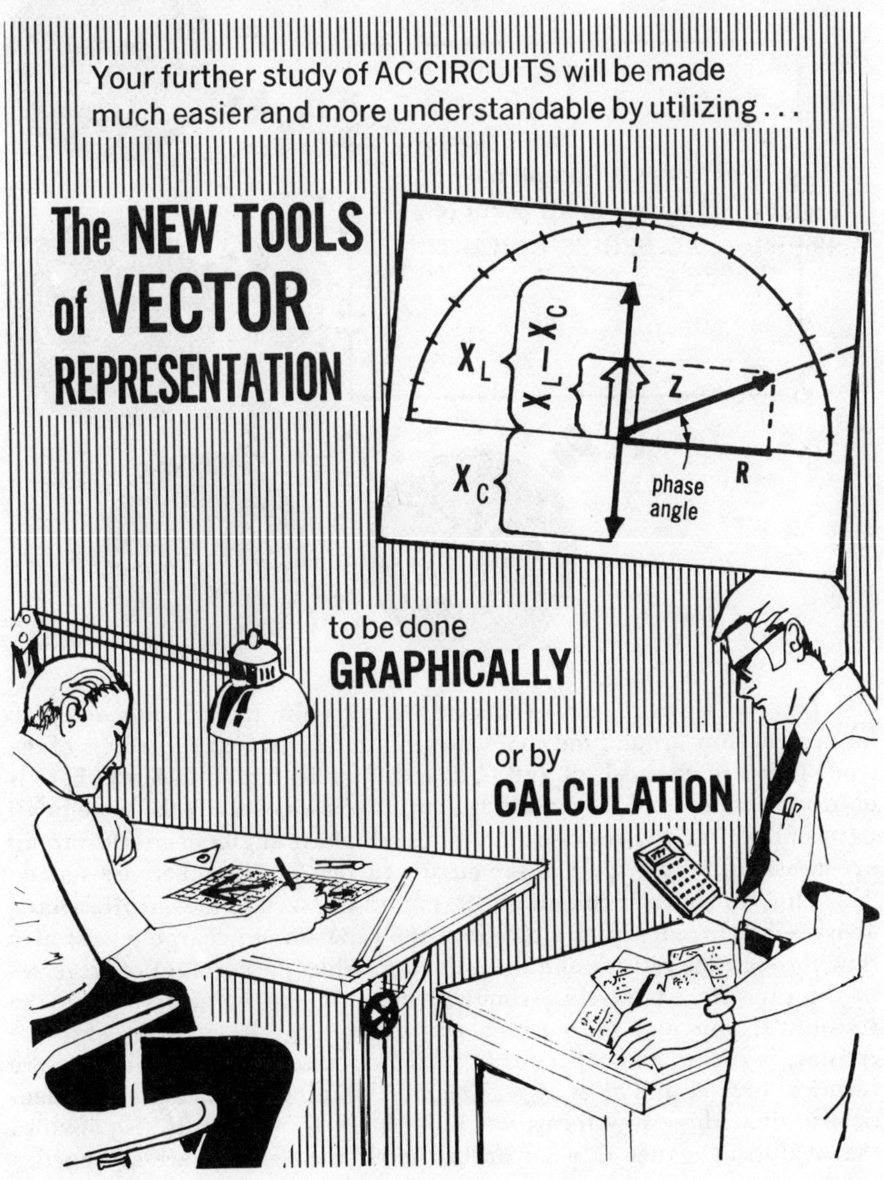

Current Flow in AC Series Circuits

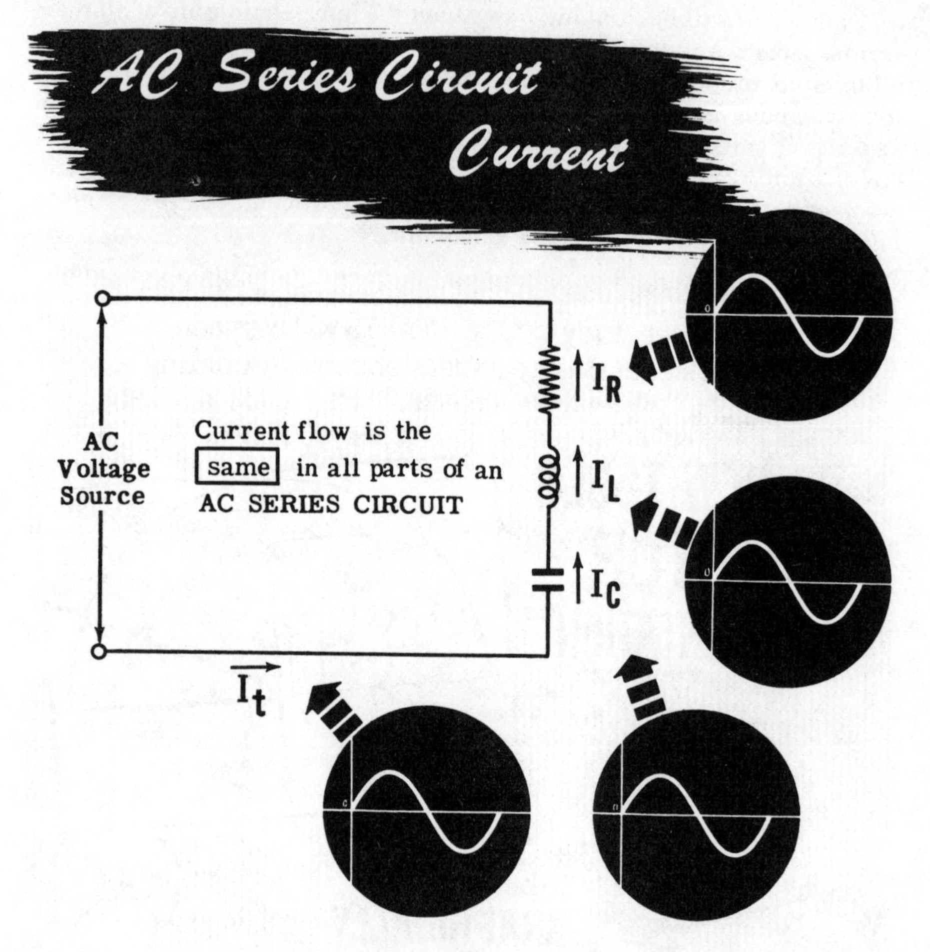

In an ac series circuit, as in a dc series circuit, there is only one path for current flow around the complete circuit. This is true regardless of the type of circuit: R and L, R and C, L and C, or R, L, and C. Since there is only one path around the circuit, the current flow is exactly the same in all parts of the circuit at any one time. Thus, all phase angles in a series circuit are measured with respect to the circuit current, unless otherwise stated.

Thus, in a circuit containing R, L, and C—such as the one illustrated above—the current which flows into the capacitor to charge it will also flow through the resistor and the inductor. When the current flow reverses in the capacitor, it reverses simultaneously in the inductor and in the resistor. If you plot the current waveforms— I_R, I_L, and I_C—for the resistor, inductor, and capacitor in such a circuit, the three waveforms are identical in *value* and *phase angle*. The total circuit current, I_t , is also identical to these three waveforms for $I_t = I_R = I_L = I_C$. It is also obvious that it doesn't matter in what order the circuit elements are arranged.

Voltages in AC Series Circuits

The total voltage, E_t, of an ac series circuit cannot be found by adding the individual voltages, E_R, E_L, and E_C, across the resistance, inductance, and capacitance of the circuit. Unlike dc circuits, these voltages cannot be added *directly* because the individual voltages across R, L, and C are *not in phase* with each other. Look below.

AC SERIES CIRCUIT VOLTAGE

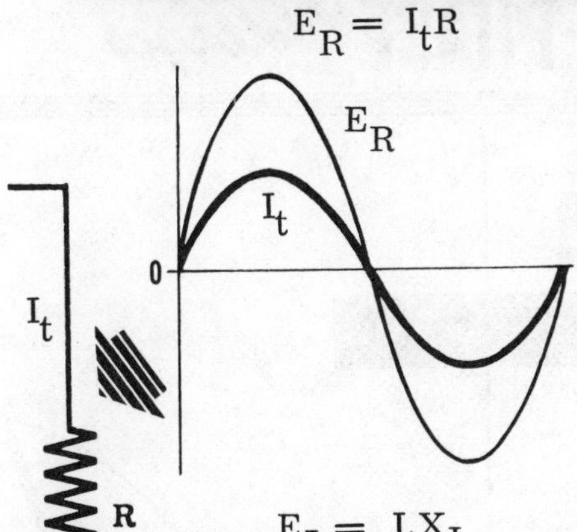

The voltage E_R across R is in phase with the circuit current, since current and voltage are in phase in pure resistive circuits.

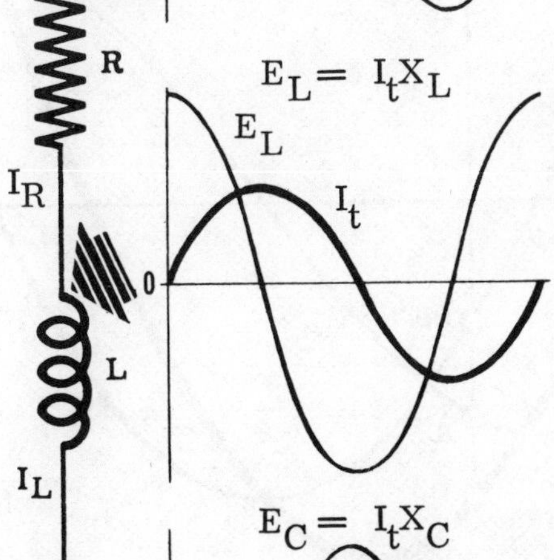

The voltage E_L across L leads the circuit current by 90 degrees, since current *lags* the voltage by 90 degrees in purely inductive circuits. Thus, E_L crosses the zero axis, going in the same direction, 90 degrees *before* the current wave.

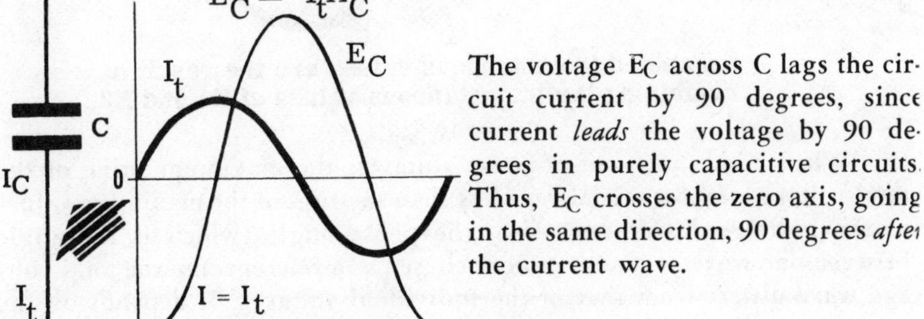

The voltage E_C across C lags the circuit current by 90 degrees, since current *leads* the voltage by 90 degrees in purely capacitive circuits. Thus, E_C crosses the zero axis, going in the same direction, 90 degrees *after* the current wave.

Voltages in AC Series Circuits (continued)

To find the total voltage in a series circuit, the instantaneous values of the individual voltages must be added together to obtain the instantaneous values of the *total voltage waveform*. Positive values are added directly, and so are negative values. The difference between the total positive and negative values for a given moment is the instantaneous value of the total voltage waveform for that instant of time. After all possible instantaneous values have been obtained, the total voltage waveform is drawn by connecting together these instantaneous values.

Combining OUT-OF-PHASE Voltages

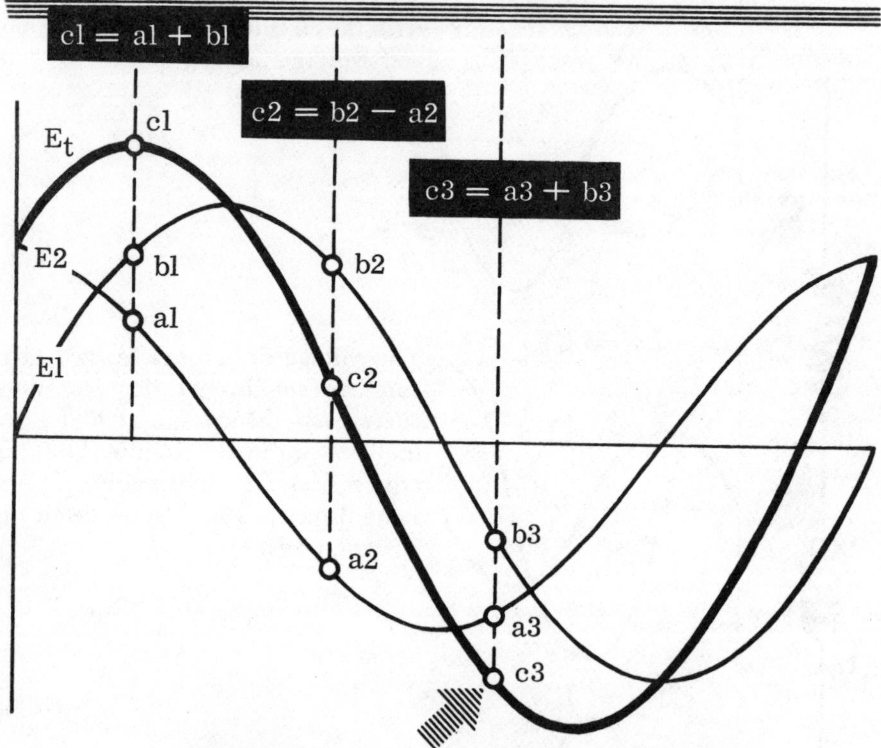

Combined instantaneous values are the result of combining the instantaneous values of E1 and E2

When combining out-of-phase voltages, the maximum value of the total voltage waveform is always less than the sum of the maximum values of the individual voltages. Also, the phase angle (which is the angle between one waveform and another chosen as a reference) of the total voltage wave differs from that of the individual voltages. It depends on the relative magnitude and phase angles of the individual voltages.

R and L Series Circuit Voltages

Suppose you consider an ac series circuit having negligible capacitance. The total circuit voltage depends on the voltage E_L across the circuit inductance and the voltage E_R across the circuit resistance. E_L leads the circuit current by 90 degrees, while E_R is in phase with the circuit current; thus E_L leads E_R by 90 degrees.

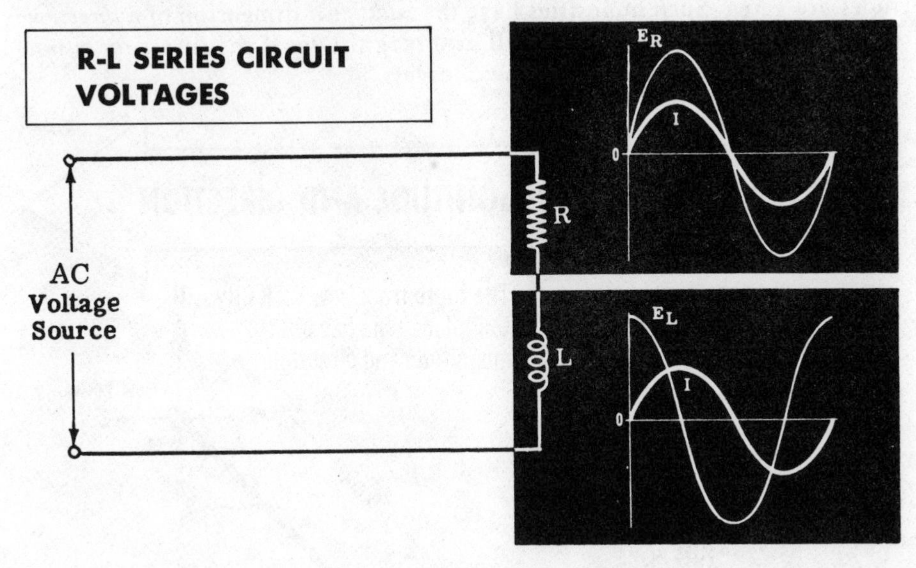

To add the voltages E_t and E_R, you can draw the two waveforms to scale and combine instantaneous values to plot the total voltage waveform. The total voltage waveform, E_t, then shows both the value and phase angle of E_t (the phase is measured with respect to the current).

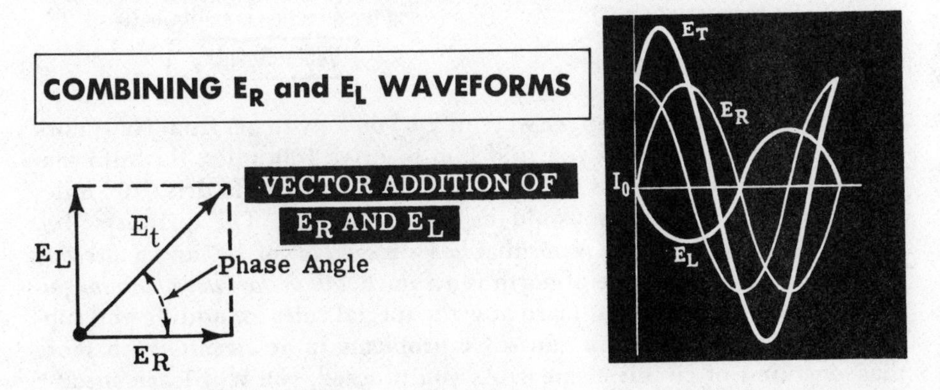

From the above, it should be clear that the resultant voltages are dependent on *both* magnitude and phase; when we add them, we must take *both* into account. To do this, we shall use *vectors* to make our work simpler with these two variables.

What a Vector Is

All physical quantities have a *basis* for specifying their size or amount. The terms "7 inches," "7 meters," "7 pounds," and "7 grams" all express the *magnitude* of physical quantities, and each is *completely* described by the number 7. Quantities that have *magnitude only* are called *scalar*. There are quantities, however, that are *not* completely described if only their magnitudes are given. Such quantities have the *additional* dimension of a *direction* with respect to a reference, as well as a magnitude. If the direction is not given, these quantities may be meaningless.

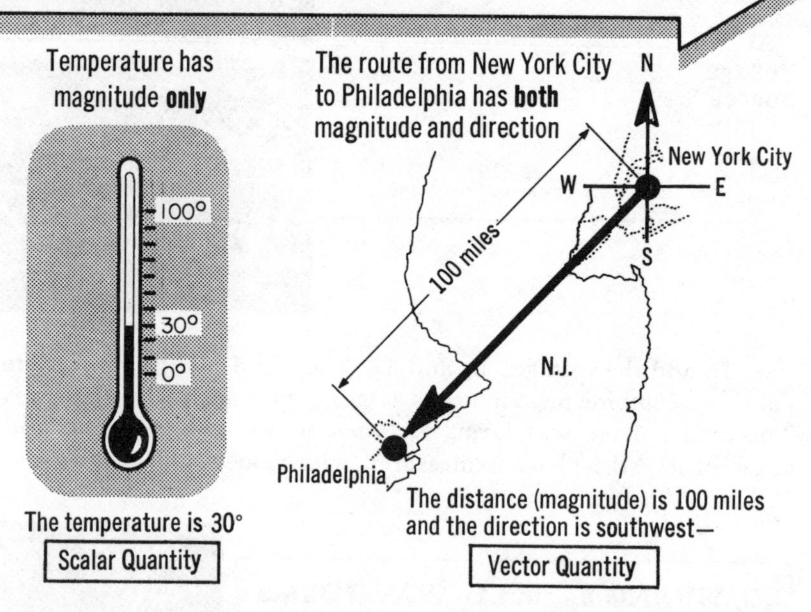

For example, if a stranger were to ask you how to get from New York City to Philadelphia, and you told him to drive 100 miles, this information would be meaningless to him. However, if you said to drive 100 miles *southwest*, your directions would be more complete. The quantity "100 miles southwest" is thus a *vector* that has a *magnitude* of *100* and a *direction* of *southwest*. The reference of north is assumed. *All vectors have both magnitude and direction.* You must learn now the special rules for adding and subtracting vectors before you can solve problems in ac circuits with more than one kind of circuit element. As you proceed, you will learn specific points about vectors that you will need to know for solving the specific ac circuits either *graphically* or *by calculation*.

A *velocity* has both *direction* and *magnitude* and is therefore a *vector* quantity. You can see how to use vectors by adding two velocities.

What a Vector Is (continued)

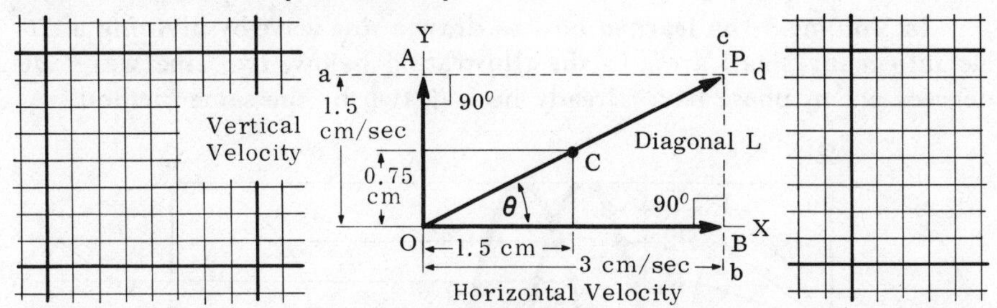

Suppose a point moves toward the top of this page from O on the line OY (OX is our reference) with a velocity of 1.5 cm/sec. One second after it starts, the point will arrive at A, which is 1.5 cm away from O. Suppose at the same time you started a point moving from O on the line OX toward B on the right-hand side of this page with a velocity of 3 cm/sec. One second after it starts, you know that it will be at B.

If the point starts at O and moves *diagonally* toward P, then it is in effect moving toward the top of the page and toward the right-hand side of the page *simultaneously.* Thus, this diagonal has *components* of velocity in *two* directions, toward A and toward B. Thus, at the end of 1 second, the point would be at P, which represents a distance of 1.5 cm vertically and 3 cm horizontally. And at the end of ½ second, the point would be at C, or 0.75 cm vertically and 1.5 cm horizontally away from O. If you had a point that had a velocity of 1.5 cm/sec toward the top of the page and simultaneously had a velocity of 3 cm/sec toward the right-hand side, you could ask, "Where will the point be at the end of 1 second?" You can represent these two velocity components as shown in the figure, by arrow OA and arrow OB. Using a scale so that OB is twice as long as OA, complete the parallelogram with lines AP and BP, drawn parallel to the axes as shown. The diagonal line OP, called the *resultant*, shows the path of the point during the 1-second interval, and P is where the point will be located at the end of 1 second. The vertical line OA and the horizontal line OB are the *vectors* representing the components or component velocities. Since the resultant diagonal line OP also has a magnitude and direction, it is thus a *resultant vector.*

The *angle* of the resultant with respect to the reference selected (line OX) gives the *direction* of the resultant. It is usually designated by the Greek letter θ (theta). Obviously, if someone gave you the location of point P, you could *resolve* the resultant into its component points by completing the parallelogram and then locating the vectors OA and OB on the drawing as the components that made up vector OP.

If you draw OA and OB carefully on graph paper and complete the parallelogram, you will find that the resultant OP is 3.35 cm long. The net velocity is thus 3.35 cm/sec. If you measure the angle θ with a protractor, you will find that the angle is about 27 degrees. As you will learn later, you can also solve vector problems analytically by using a hand calculator.

Vector Graphic Representation of AC Voltages and Currents

In Volume 3 you learned how to draw a sine wave by dividing a circle into many small arcs. In the illustration below, two sine waves 90 degrees out of phase have already been drawn by the same method.

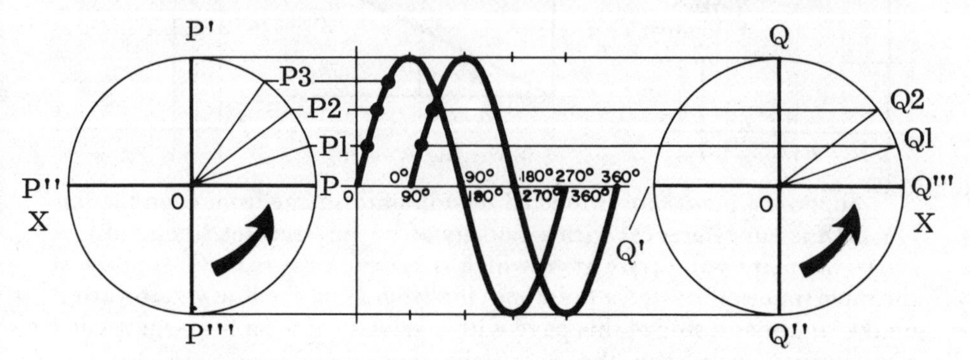

If you imagine the points P and Q to be rotating counterclockwise at the same speed on the two circles, you can see that the vertical distances of P and Q above the axis XX give the instantaneous values of the sine waves. As P and Q rotate through 360 degrees (one revolution), the sine waves complete one cycle. However, as the sine waves are 90 degrees out of phase, P reaches P′ when Q‴ has just arrived at the axis XX (the starting reference). Q starts to rotate and when Q reaches Q′, P′ has reached P″, and so on. Thus, to illustrate the relationship between the two sine waves (which could be, for example, the two voltages in a resistance and an inductor in series measured with respect to the current), you can use lines OP and OQ, provided that you draw them so that the angle between them is *equal to the phase angle between the sine waves.* Then OP and OQ are also vectors.

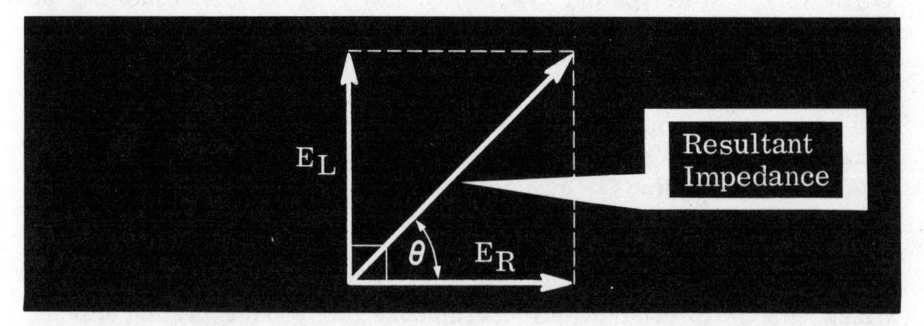

You have already learned that in an ac series circuit containing both inductance and resistance, the voltage across the inductance leads the voltage across the resistance by 90 degrees. You can, therefore, represent the voltages by vectors drawn at right angles to each other, as illustrated. The vectors are presumed to be rotating in a counterclockwise direction, so the fact that voltage across the inductor leads the voltage across the resistance is, in this case, represented by drawing the inductor voltage vector *vertically* and the resistance voltage vector *horizontally.*

Vector Graphic Representation of AC Voltages and Currents (continued)

If two out-of-phase voltages or currents are to be added together, you can represent them by vectors and find their resultant in the manner described on page 4-9. The lengths of the vectors are made proportional to the magnitude of the two quantities, and the angle between the vectors is made equal to the phase angle (θ) between them.

You can find the total voltage E_t in the R and L series circuit shown on page 4-7 by representing E_R and E_L by vectors and by finding the resultant using the method just described.

Draw a vector horizontally to represent E_R (the voltage across the resistance). E_R is in phase with the current, which is used as the reference vector. As the voltage E_L (the voltage across the inductor) leads the current by 90 degrees, draw the vector representing E_L at 90 degrees to E_R, ahead of E_R. (Remember that the vectors are rotating in a *counterclockwise* direction.) The resultant, E_t, is then obtained by completing the parallelogram. Draw lines parallel to E_L and E_R and join their crossing point at P (intersect at P).

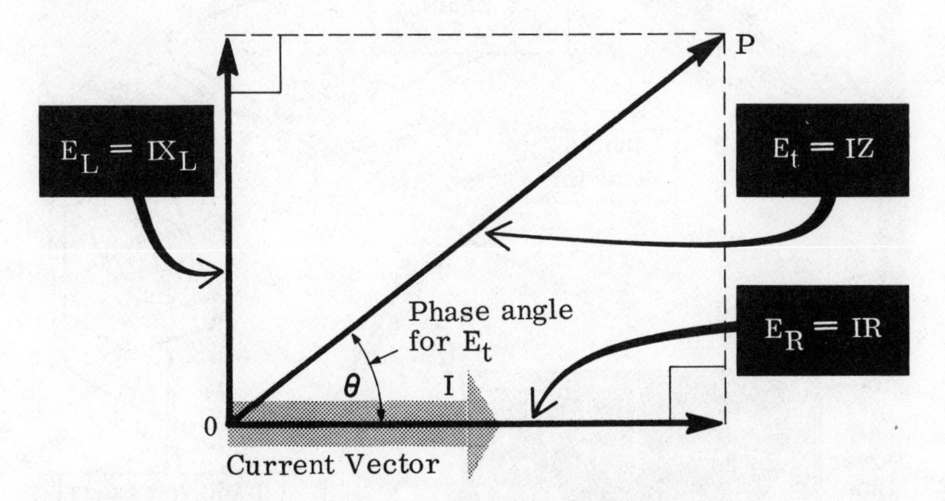

As E_R is in phase with the current through the circuit, the phase angle of the total voltage is the angle between E_R and E_t. In the case of an R and L series circuit, the total voltage leads the current.

If you now apply Ohm's law to the ac circuits you are considering, you will see that:

$$E_R = IR$$
$$E_L = IX_L \text{ (where } X_L \text{ is the inductive reactance)}$$

Z is the total opposition to current flow in the circuit and is called the *impedance.* You will learn about impedance on pages 4-13 to 4-22.

$$E_t = IZ$$

Experiment/Application—R and L Series Circuit Voltages

To verify the relationship of the various voltages in an R and L series circuit, you could connect a 1,500-ohm resistor and a 5-henry inductor to form an R and L series circuit (see illustration below). With the switch closed, individual voltage readings are taken across the inductor and resistor. Also, the total voltage across the series combination of the resistor and inductor is measured. Notice that if the measured voltages across the inductor and resistor are added directly, the result is greater than the total voltage measured across the two in series.

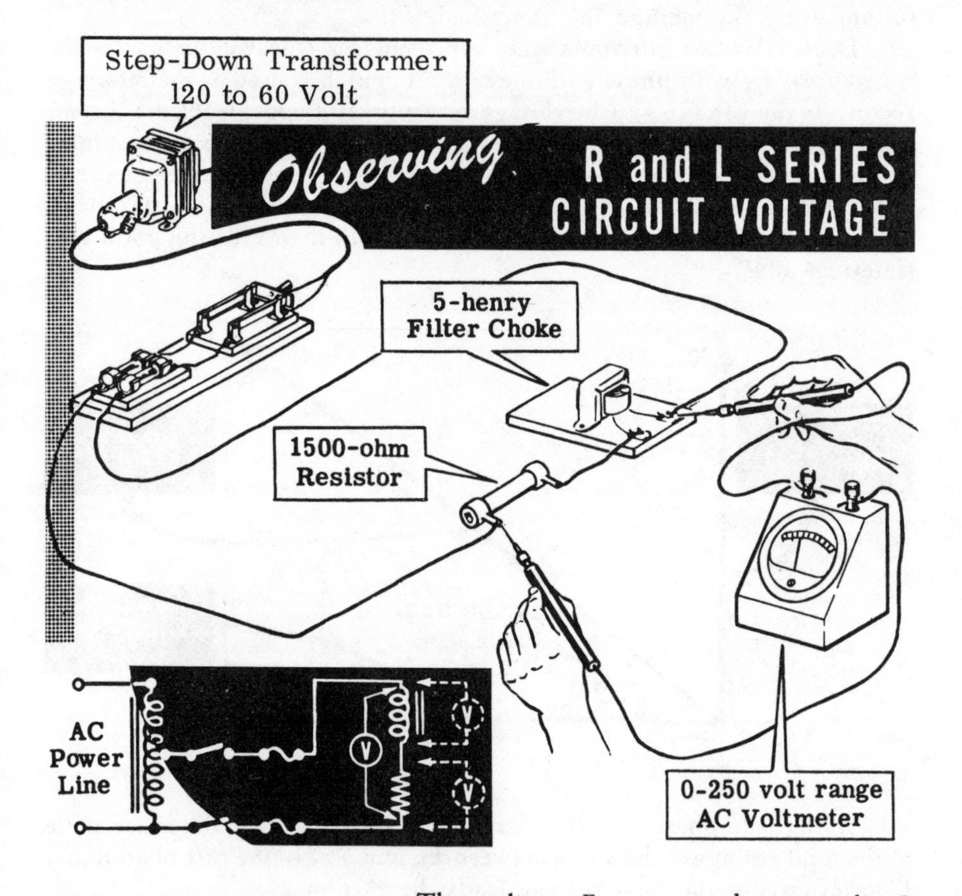

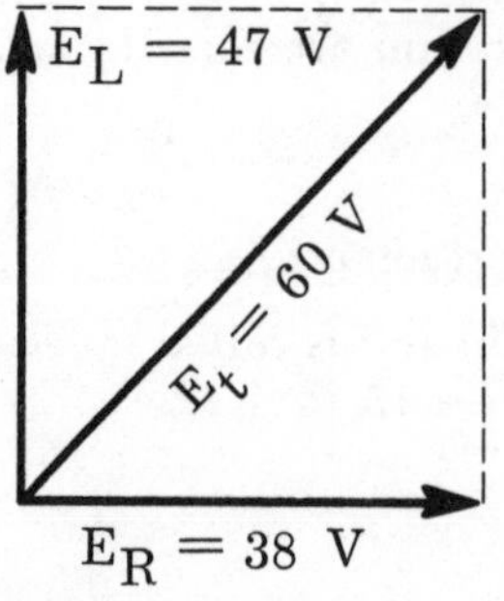

The voltage E_L measured across the 5-henry inductor is approximately 47 volts, and E_R measured across the 1,500-ohm resistor is about 38 volts. When added directly, the voltages E_L and E_R total approximately 85 volts, but the actual measured voltage across the resistor and inductor in series is only about 60 volts. Using vectors to combine E_R and E_L, you see that the result is about 60 volts—the actual total circuit voltage *as measured.*

R and L Series Circuit Impedance

As you know, when resistance and inductance are connected in series, the impedance (Z) is *not* found by adding these two values directly. Inductive reactance causes voltage to lead the current by 90 degrees, whereas for pure resistance the voltage and current are in phase. Thus, the effect of inductive reactance, as opposed to the effect of resistance, is shown by drawing two vectors to represent R and X_L at right angles to one another.

Look again at the voltage vector diagram for the R and L circuit on page 4-11. You will see that the lengths of the voltage vectors are proportional to the magnitudes of the voltages, E_L, E_R, and E_t.

Since $E_L = IX_L$, $E_R = IR$, and $E_t = IZ$, the vectors also represent, in their true proportions and relationship, X_L, R, and Z.

For example, suppose that the series circuit consists of 200 ohms of resistance in series with 200 ohms of inductive reactance at the frequency of the ac voltage source. The total impedance is *not* 400 ohms; it is approximately 283 ohms. *This value of impedance must be obtained by vector addition.*

To add 200 ohms of resistance and 200 ohms of inductive reactance, a horizontal line is drawn to represent the 200 ohms of resistance. The end of this line, which is to the left, is the reference point; the right end of the line is marked with an arrow to indicate its direction. To represent the inductive reactance, a vertical line is drawn upward from the reference point. Since X_L and R are each 200 ohms, the horizontal and vertical lines are equal in length.

In a series circuit, the resistance vector is usually drawn horizontally and used as a reference for other vectors in the same diagram.

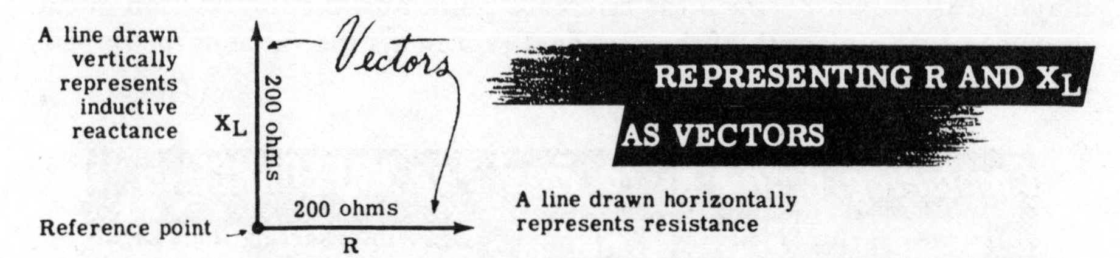

A parallelogram is completed as shown in the right-hand diagram below. The diagonal of this parallelogram represents the impedance Z, the total opposition to current flow of the R and L combination. Using the same scale of measure, this value is found to be 283 ohms at a phase angle of 45 degrees.

COMBINING VECTORS R AND X_L TO FIND THE IMPEDANCE

R and L Series Circuit Impedance (continued)

If the resistance and inductance values of a series circuit containing R and L are known, the impedance can be found by means of vectors. Suppose that the resistance is 180 ohms, the inductance is 400 mH, and the frequency of the applied ac voltage is 60 Hz. First, the inductive reactance, X_L, is computed in ohms by using the formula $X_L = 2\pi fL$. $X_L = 2 \times \pi \times 60 \times 0.4 = 150$ ohms. Then the vectors representing R and X_L are drawn to scale on graph paper. A common reference point is used with the resistance vector R drawn horizontally to the right, and the inductive reactance vector X_L is drawn upward, perpendicular to the resistance vector.

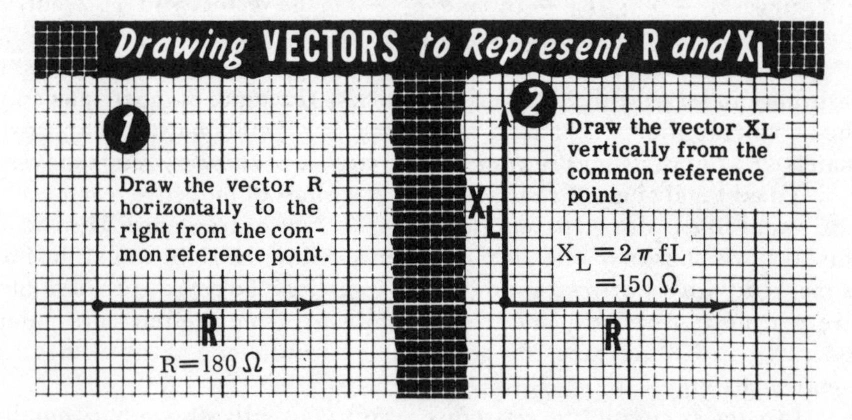

Next, the parallelogram is completed using dotted lines, and the diagonal is drawn between the reference point and the intersection of the dotted lines. The length of the diagonal represents the value of the impedance Z in ohms.

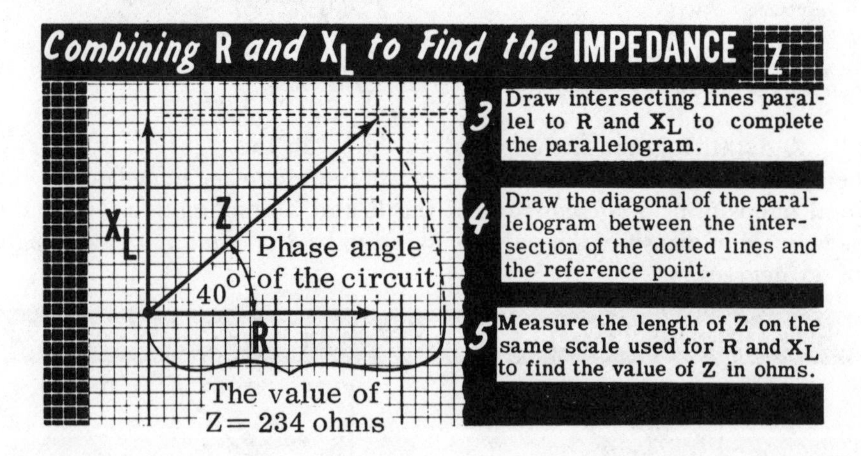

In addition to showing the value of the circuit impedance, the vector solution also shows the phase angle between the circuit current and voltage. The angle between the impedance vector Z and the resistance vector R

R and L Series Circuit Impedance (continued)

is the *phase angle* of the circuit in degrees. This is the angle between the circuit voltage and current and represents a current lag of 40 degrees (or voltage lead of 40 degrees).

Ohm's law for ac circuits may also be used to find the impedance Z for a series circuit. In applying Ohm's law to an ac circuit, Z is substituted for R in the formula. Thus, the impedance, Z, is equal to the applied voltage, E, divided by the circuit current, I. For example, if the circuit voltage is 120 volts and the current 0.5 ampere, the impedance Z is 240 ohms for a frequency of 60 Hz.

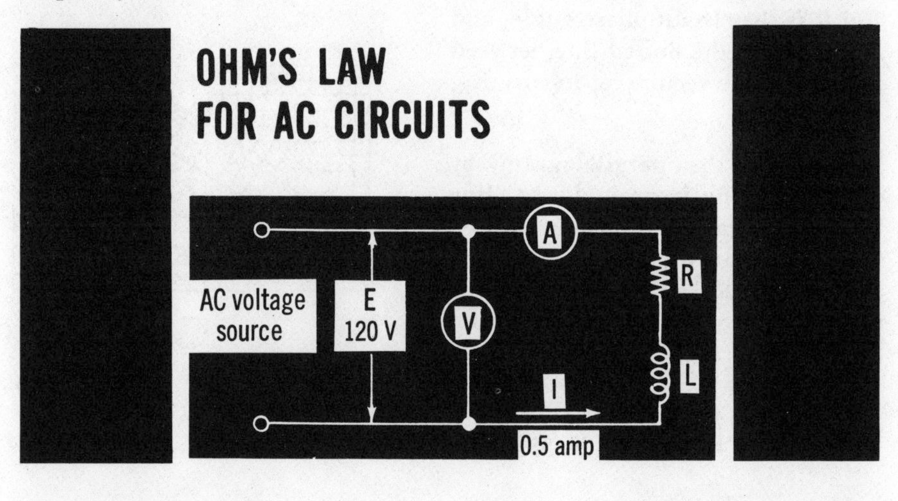

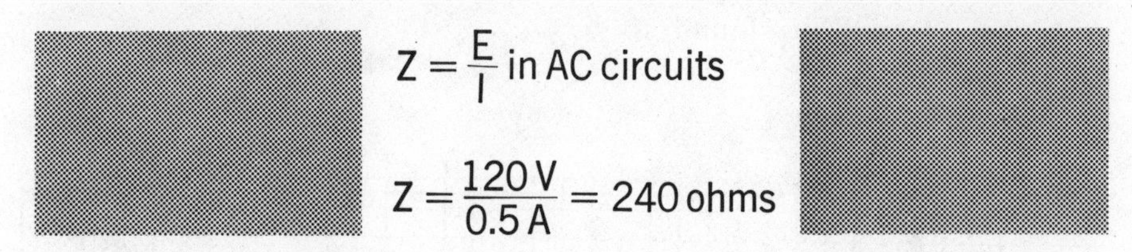

If the impedance of a circuit is found by applying Ohm's law for ac, and the value of R is known and X_L is unknown, the phase angle and the value of X_L may be determined graphically by using vectors. If the resistance in the circuit above is known to be 200 ohms, the vector solution is:

1. Since the resistance is known to be 200 ohms, the resistance vector is drawn horizontally from the reference point. At the end of the resistance vector, a dotted line is drawn perpendicular to it.

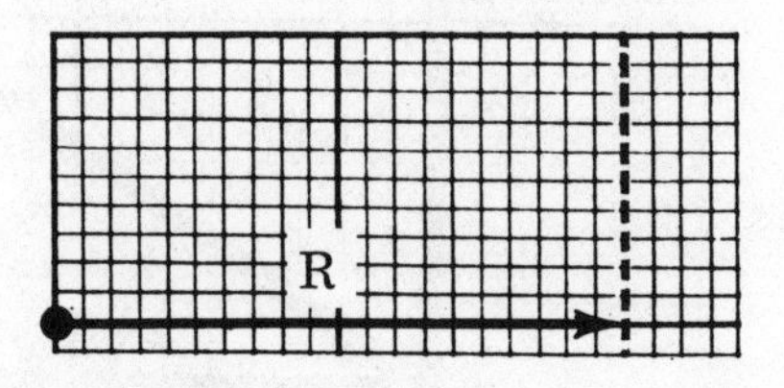

R and L Series Circuit Impedance (continued)

2. Using a straight edge (ruler) marked to indicate the length of the impedance vector, Z, find the point on the perpendicular dotted line which is exactly the length of the impedance vector from the reference point. Draw the impedance vector between that point and the zero position. The angle between the vectors Z and R is the circuit phase angle, and the length of the dotted line between the ends of the vectors represents X_L.

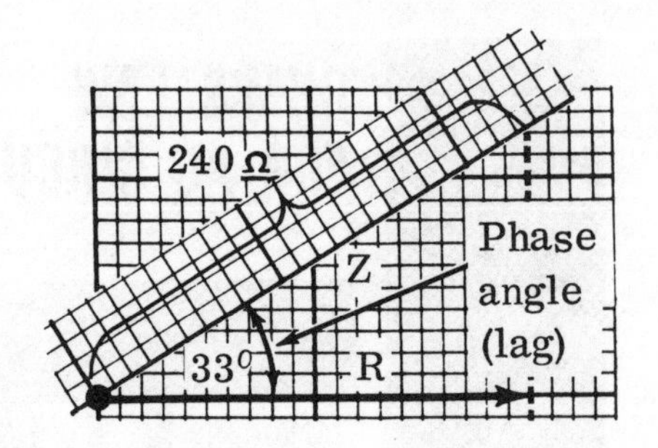

3. Complete the parallelogram by drawing a horizontal dotted line between the end of the vector Z and a vertical line drawn up from the reference point. X_L is this vertical line, and its length can be read by using the same scale. In the example shown, X_L = 133 ohms. And since $X_L = 2\pi fL$, L must equal $X_L/2\pi f$. Therefore, L = 0.353 H, or 353 mH.

4. Measure the phase angle with a protractor. It will be found to be about 33 degrees.

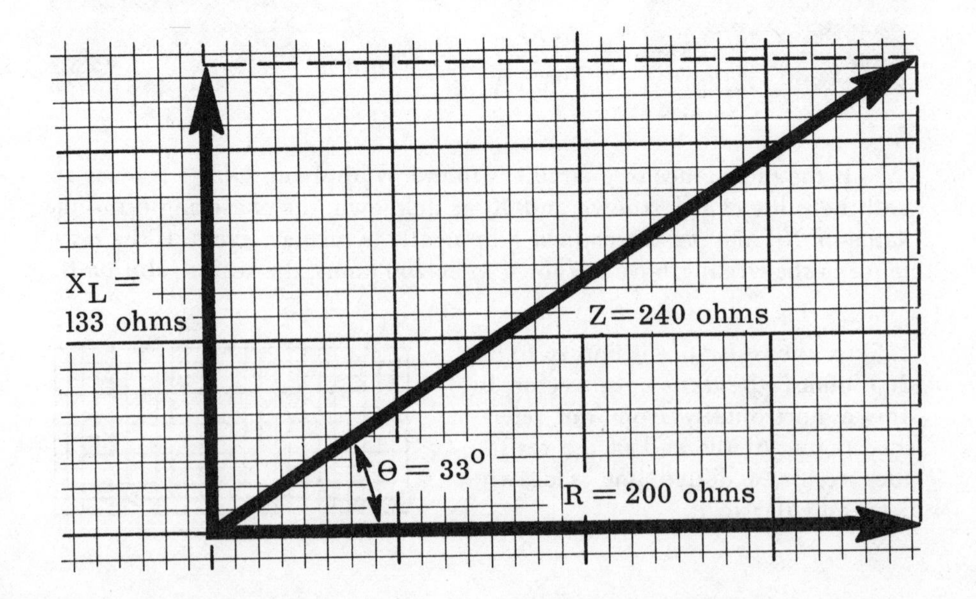

R and L Series Circuit Impedance (continued)

If the impedance and inductance are known, but the resistance is not, both the phase angle and resistance may be found by using vectors. For example, the impedance is calculated to be 300 ohms by measuring the current and voltage and applying Ohm's law for ac. If the circuit inductance is 0.5 henry, the vector solution is as follows:

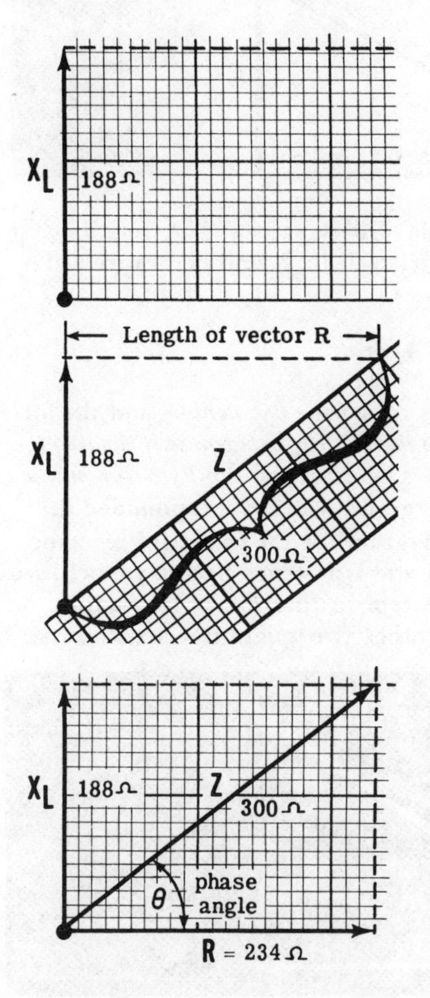

1. First the inductive reactance is computed by using the formula $X_L = 2\pi fL$. If the frequency is 60 Hz, then X_L is 188 ohms ($X_L = 6.28 \times 60 \times 0.5 = 188$ ohms). Draw the vector X_L vertically from the reference point. At the end of this vector draw a horizontal dotted line perpendicular to vector X_L.

2. Using a straight edge marked to indicate the length of the impedance vector, find the point on the horizontal dotted line which is exactly the length of the impedance from the reference point. Draw the impedance vector Z between that point and the reference point. The distance between the ends of the vectors X_L and Z represents the length of the resistance vector R.

3. Draw the vector R horizontally from the zero position and complete the parallelogram. The angle between R and Z is the phase angle of the circuit in degrees. The length of the vector R represents the resistance in ohms and is found to be 234 ohms.

4. Measure the phase angle with a protractor. It will be found to be 39 degrees.

You have already learned how to solve series circuits graphically and to calculate (by Ohm's law) the impedance of a series circuit which is composed of an inductor and a resistor connected to an ac voltage source. You will now learn how to calculate the impedance of such a circuit *without* the use of Ohm's law and *without* making measurements. A hand calculator is very useful for these calculations, and you will find that they are much more rapid and accurate than graphical methods.

R and L Series Circuit Impedance (continued)

To make use of them, you will have to know something about right triangles—that is, a triangle with a right angle. In any right-angled triangle, the lengths of the three sides bear a definite relationship to one another. Consider the triangle below:

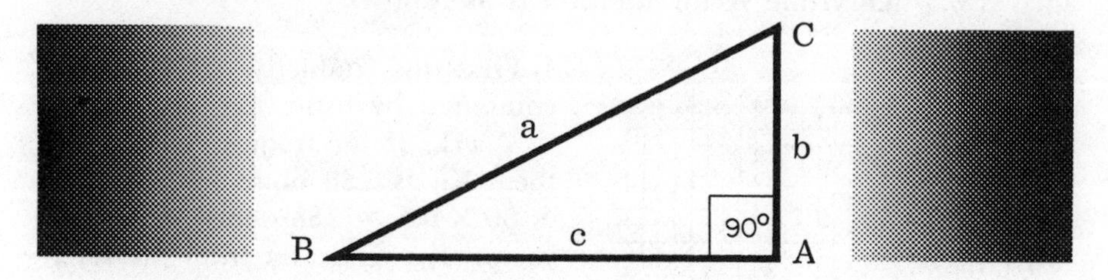

The angle BAC is a right angle (90 degrees). The relationship between the lengths of the sides of this triangle (a, b, and c) is expressed by the formula:

$$a^2 = b^2 + c^2$$

Side *a*, opposite the right angle, is called the *hypotenuse,* and the formula can be expressed in words as *the square of the hypotenuse is equal to the sum of the squares of the other two sides.* This is called the *Pythagorean theorem,* after the Greek philosopher, Pythagoras, who first propounded it.

If you look again at the vector diagram for the inductive reactance, resistance, and total impedance in an R and L ac series circuit, you will see that you can use the Pythagorean theorem to find either impedance, or reactance, or resistance, provided the other two quantities are known.

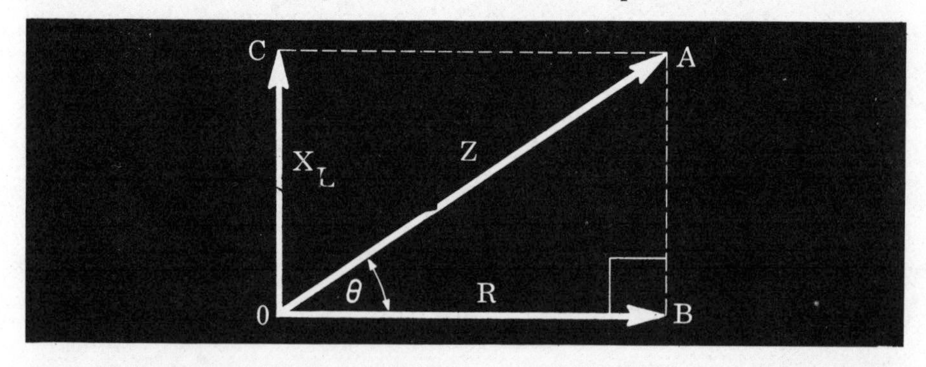

You can see that OAB (above) is a right-angled triangle in which Z is the hypotenuse, R (OB) one side, and X_L (AB= OC) the other. Therefore,

$$Z^2 = R^2 + X_L{}^2$$

and

$$Z = \sqrt{R^2 + X_L{}^2}$$

R and L Series Circuit Impedance (continued)

The two angles of a right triangle (other than the right angle itself) are related to the lengths of the sides by the trigonometric relationships called the *sine, cosine,* and *tangent*, abbreviated *sin, cos*, and *tan*, respectively. If the magnitude of an angle is known, the value of the sin, cos, or tan of the angle can be found in the tables of trigonometric functions at the back of this volume. Obviously, you can reverse the procedure and find the angle, if you know the value for the sin, cos, or tan, by looking up this value in the same tables.

$$\text{Sine of either angle} = \frac{\text{Length of side } \textit{opposite} \text{ angle}}{\text{Length of hypotenuse}}$$

$$\text{Cosine of either angle} = \frac{\text{Length of side } \textit{adjacent} \text{ to angle}}{\text{Length of hypotenuse}}$$

$$\text{Tangent of either angle} = \frac{\text{Length of side } \textit{opposite} \text{ angle}}{\text{Length of side } \textit{adjacent} \text{ to angle}}$$

Examples

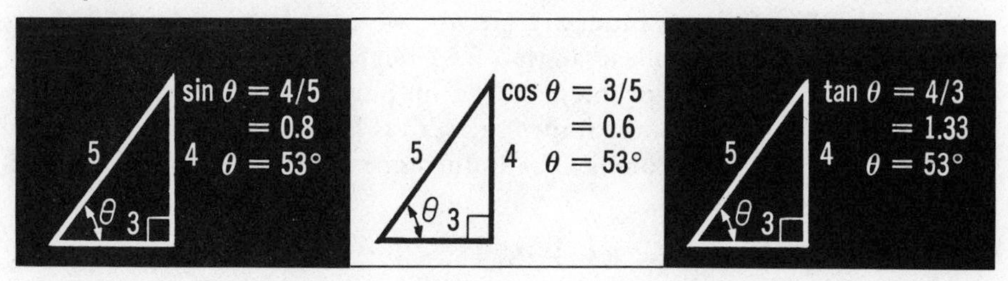

When the length of the hypotenuse and the angle between it and one of the other sides are known, the length of the other two sides can be found using the trigonometric relationships of sin or cos.

Example

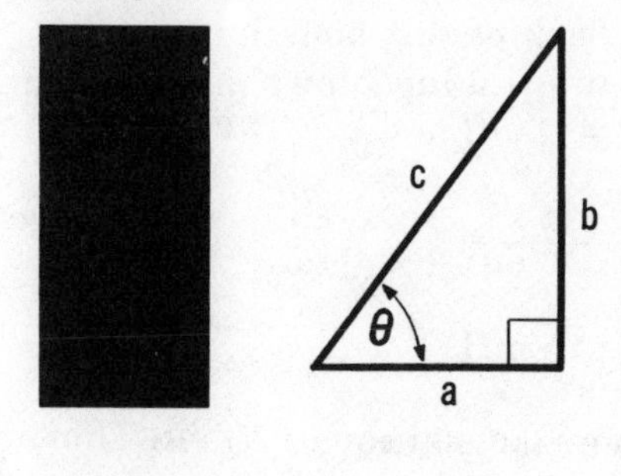

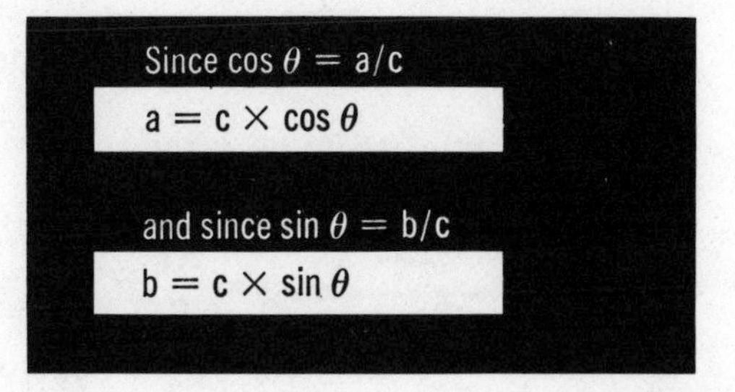

R and L Series Circuit Impedance (continued)

Suppose that in an ac series circuit the value of X_L and R are known to be 4 ohms and 3 ohms, respectively. Using the formula for the impedance Z from the previous pages:

$$Z = \sqrt{R^2 + X_L^2}$$
$$Z = \sqrt{3^2 + 4^2}$$
$$= \sqrt{9 + 16}$$
$$= \sqrt{25}$$
$$Z = 5 \text{ ohms}$$

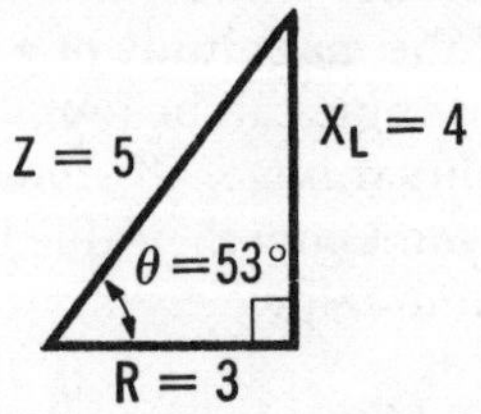

Z = 5
$X_L = 4$
$\theta = 53°$
R = 3

Phase angle θ is the angle between Z and R, and the tangent of that angle is X/R. Thus θ is the angle whose tangent is X/R, which is written algebraically as $\theta = \tan^{-1} X/R$ ($\tan^{-1}$ is read as "the angle whose tangent is"). Since we know that $X/R = 4/3 = 1.333$, $\theta = \tan^{-1} 1.333$.

The angle can now be found, by consulting the trigonometric tables at the back of this volume, to be 53 degrees, approximately.

As the circuit is an inductive circuit, we now know that the voltage leads the current by a phase angle of 53 degrees.

Consider again the problem solved on page 4-15 by drawing vectors to scale. In that problem, the impedance (Z) is known to be 240 ohms, and the resistance (R) is 200 ohms. The inductance and phase angle of the circuit remain to be found.

$$Z^2 = R^2 + X_L^2$$
$$X_L^2 = Z^2 - R^2$$
$$X_L = \sqrt{Z^2 - R^2} = \sqrt{240^2 - 200^2}$$
$$= \sqrt{57{,}600 - 40{,}000} = \sqrt{17{,}600}$$
$$X_L = 133 \text{ ohms, approximately}$$

(When doing calculations of this sort, you can most easily find the squares and square roots you want by looking them up in a set of mathematical tables or using the tables at the back of this book.)

The inductance can now be found by substituting known quantities in the formula for inductive reactance, $X_L = 2\pi fL$:

$$L = \frac{X_L}{2\pi f} = \frac{133}{6.28 \times 60}$$

$$L = 0.353 \text{ H or } 353 \text{ mH}$$

The phase angle, we know, is the angle whose tangent equals X_L/R. Thus,

$$\tan\theta = \frac{X_L}{R} = \frac{133}{200} = 0.665$$

$$\theta = 33.5° \text{ approximately}$$

Experiment/Application—Impedance in R and L Series Circuits

You can verify the methods of calculating impedance described in the preceding pages by first calculating the impedance of a given circuit, using both methods, and then measuring it in an actual circuit.

Consider an R and L circuit consisting of a 5-henry inductor in series with a 2,000-ohm resistor. First, calculate the inductive reactance, X_L:

$$X_L = 2\pi fL = 2\pi \times 60 \times 5 = 1{,}885 \text{ ohms}$$

Now, using the formula on page 4-18, calculate the impedance, Z:

$$Z = \sqrt{R^2 + X_L^2} = \sqrt{2{,}000^2 + 1{,}885^2} = 2{,}750 \text{ ohms, approximately}$$

The phase angle is calculated as $\theta = \tan^{-1} (X/R) = \tan^{-1} (1{,}885/2{,}000) = 43$ degrees.

A similar value for Z and θ will be found if a vector diagram is drawn to scale, and Z, the resultant, and θ, the phase angle, are measured.

Now connect a 2,000-ohm resistor, a 5-henry inductor, a switch, a fuse and a 0–50 mA ac milliammeter in series across the ac power source through a step-down transformer (120 to 60 volts) to form a series R and L circuit. A 0–150 volt ac voltmeter is connected across the transformer secondary to measure the circuit voltage. With the switch closed, you see that the voltmeter reads approximately 60 volts and the milliammeter reads about 22 mA. The Ohm's law value of Z is about 2,750 ohms ($60 \div 0.022 \approx 2{,}750$), and you see that the two methods of finding Z result in approximately equal values for Z. (The meter readings you observe will vary somewhat from those given because of variations in line voltage and meter accuracy and in the rating of the resistors and inductors used. Thus, the values of Z obtained in practice will always vary slightly from the values obtained by calculation.)

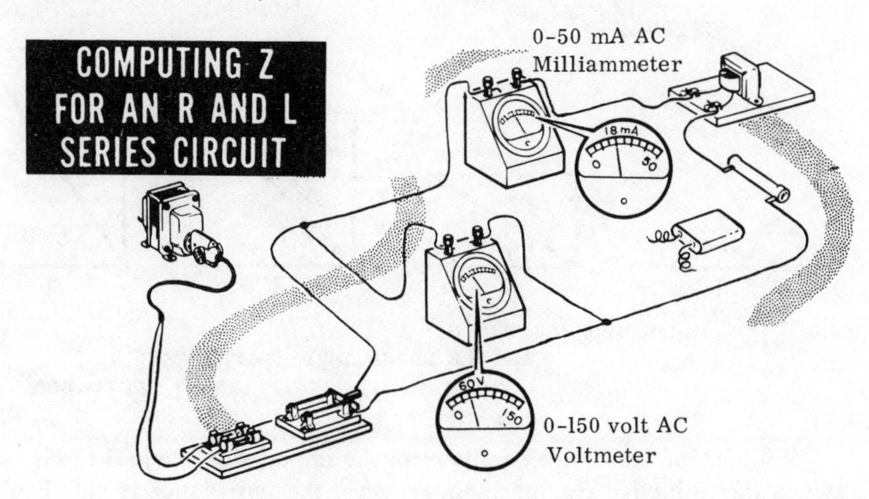

R and L Series Circuits Impedance Variation

The impedance of a series circuit containing only resistance and inductance is determined by the vector addition of the resistance and the inductive reactance. If a given value of inductive reactance is used, the impedance varies as shown below when the resistance value is changed.

How Impedance Varies

WHEN X_L IS A FIXED VALUE AND R IS VARIED

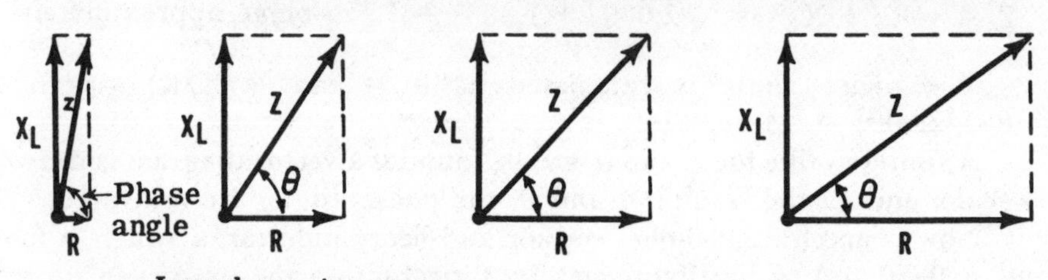

Impedance increases
and the phase angle decreases
as R increases

If a given value of resistance is used, the impedance, in its turn, varies as shown below when the inductive reactance is changed:

How Impedance Varies

WHEN R IS A FIXED VALUE AND X_L IS VARIED

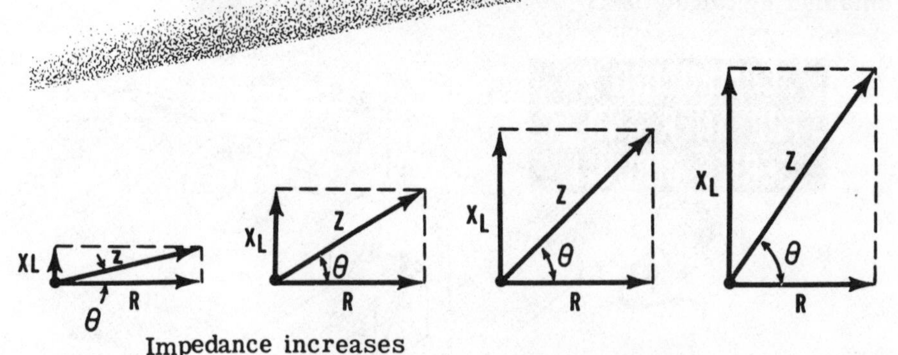

Impedance increases
and the phase angle increases
as X_L increases

When the resistance equals zero, the impedance is equal to X_L, and when the inductive reactance equals zero, the impedance is equal to R.

R and C Series Circuit Voltages

If an ac series circuit consists of only R and C, the total voltage is found by vectorially combining E_R, the voltage across the resistance, and E_C, the voltage across the capacitance. E_R is in phase with the circuit current, while E_C lags the circuit current by 90 degrees; thus, E_C lags by 90 degrees. The two voltages may be combined either graphically or by calculation as you did for R-L circuits.

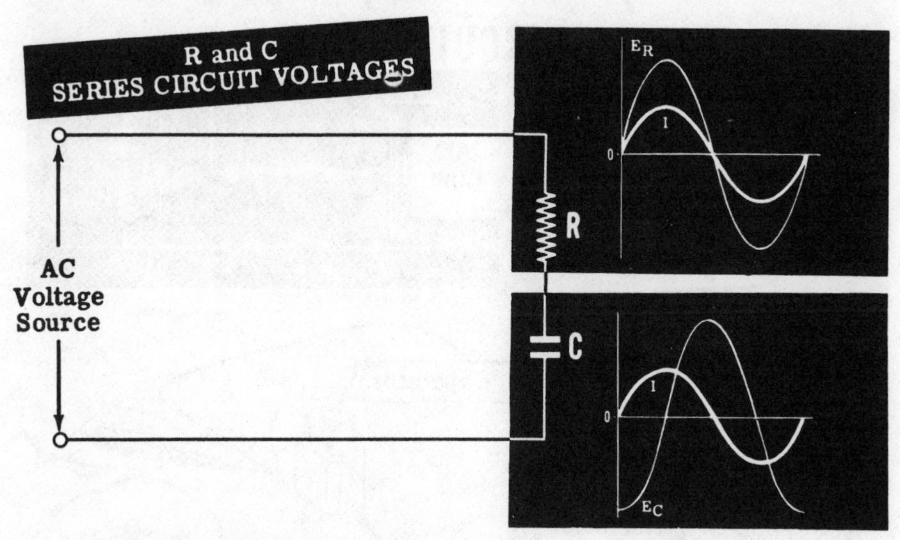

As the voltage across C lags the current in the circuit, the vector representing E_C is drawn downward from the horizontal vector representing E_R. Similarly, in the impedance vector diagram, X_C is drawn downward from R.

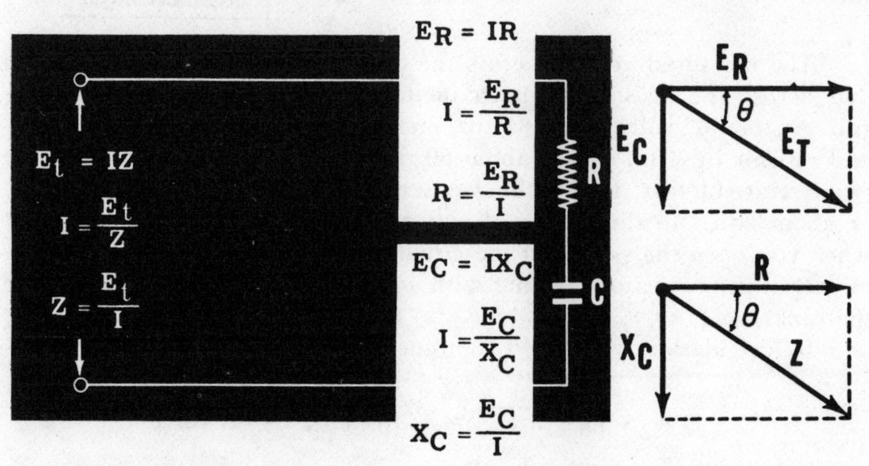

Experiment/Application—R and C Series Circuit Voltages

If you repeat the experiment/application described on page 4-12, using a 1-μF capacitor in series with a 1,500-ohm resistor, again you see that the sum of the voltages across the capacitor and resistor is greater than the actual measured total voltage.

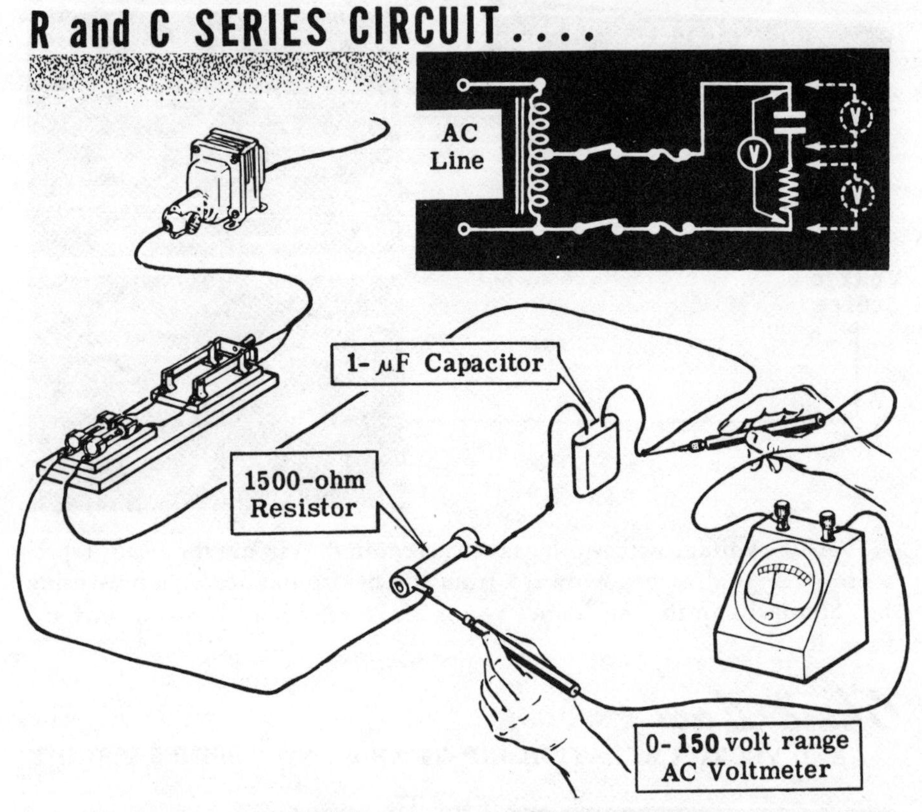

The measured voltage across the capacitor is about 52 volts, while that across the resistor is approximately 30 volts. When added, these voltages total 82 volts, but the actual measured voltage across the capacitor and resistor in series is only about 60 volts. Using vectors to combine the two circuit voltages, E_R and E_C, you see that the answer is about 60 volts, or about equal to the measured voltage of the circuit. (Remember that when you open the switch in the circuit, you must always short the terminals of the capacitor together with a screwdriver in order to discharge the capacitor.)

By calculation, you could combine the voltages as:

$$E_t = \sqrt{E_R^2 + E_C^2} = \sqrt{30^2 + 52^2} = 60 \text{ volts}$$

R and C Series Circuit Impedance

If an ac series circuit consists of resistance and capacitance in series, the total opposition to current flow (impedance) is due to two factors, resistance and capacitive reactance. The action of capacitive reactance causes the current in a capacitive circuit to lead the voltage (voltage lags current) by 90 degrees, so that the effect of capacitive reactance is at right angles to the effect of resistance. But, though the effects of inductive and capacitive reactance are both at right angles to the effect of resistance, their effects are exactly *opposite*—inductive reactance causing current to lag and capacitive reactance causing it to lead the voltage. Thus, the vector X_C, representing capacitive reactance, is still drawn perpendicular to the resistance vector, but drawn *down* rather than *up* from the zero position.

The impedance (Z) of a series circuit containing R and C is found in the same manner as the impedance of an R and L series circuit. Suppose in your R and C series circuit that R equals 200 ohms and X_C equals 200 ohms. To find the impedance, the resistance vector R is drawn horizontally from a reference point. Then a vector of equal length is drawn downward from the reference point at right angles to the vector R. This vector X_C, which represents the capacitive reactance, is equal in length to R since both R and X_C equal 200 ohms.

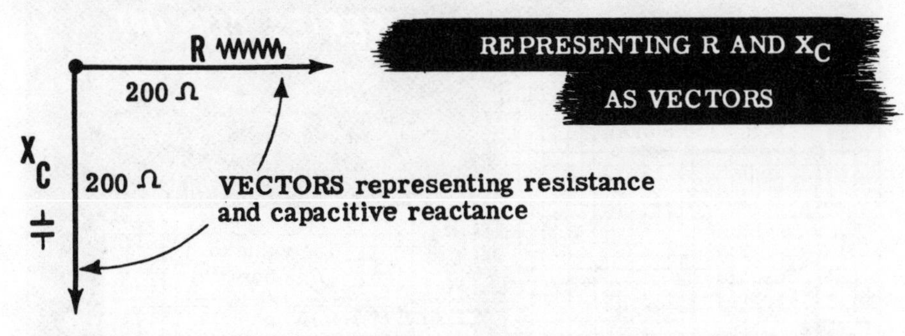

REPRESENTING R AND X_C AS VECTORS

To complete the vector solution, the parallelogram is completed and a diagonal drawn from the reference point. This diagonal is the vector Z and represents the impedance in ohms (283 ohms). The angle between the vectors R and Z is the phase angle (θ) of the circuit, indicating the amount in degrees by which the current leads the voltage.

COMBINING VECTORS R AND X_C TO FIND THE IMPEDANCE

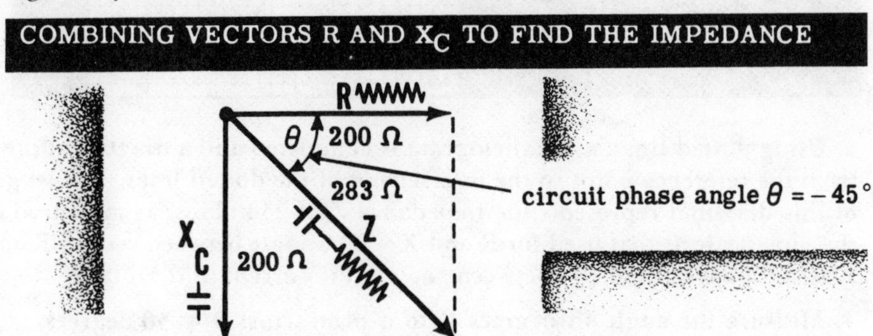

R and C Series Circuit Impedance (continued)

R and C series circuit impedances can be found by using the vector solution, by calculation, or by application of Ohm's law. To find the impedance Z (for example, in a series circuit containing 150 ohms resistance and 15 μF of capacitance) by using a vector solution, you should proceed as follows:

1. Compute the value of X_C by using the formula $X_C = 1/2\pi fC$. In this formula, 2π is a constant equal to 6.28, f is the frequency in Hz, and C is the capacitance in farads. For example, if C = 15 μF, $X_C = 1/6.28 \times 60 \times 0.000015$ = 177 ohms, or approximately 180.

2. Given that R = 150 ohms, draw vectors R and X_C to scale on graph paper, using a common reference point for the two vectors. R is drawn horizontally to the right from the reference point, and X_C is drawn downward from the reference point and perpendicular to the resistance vector R.

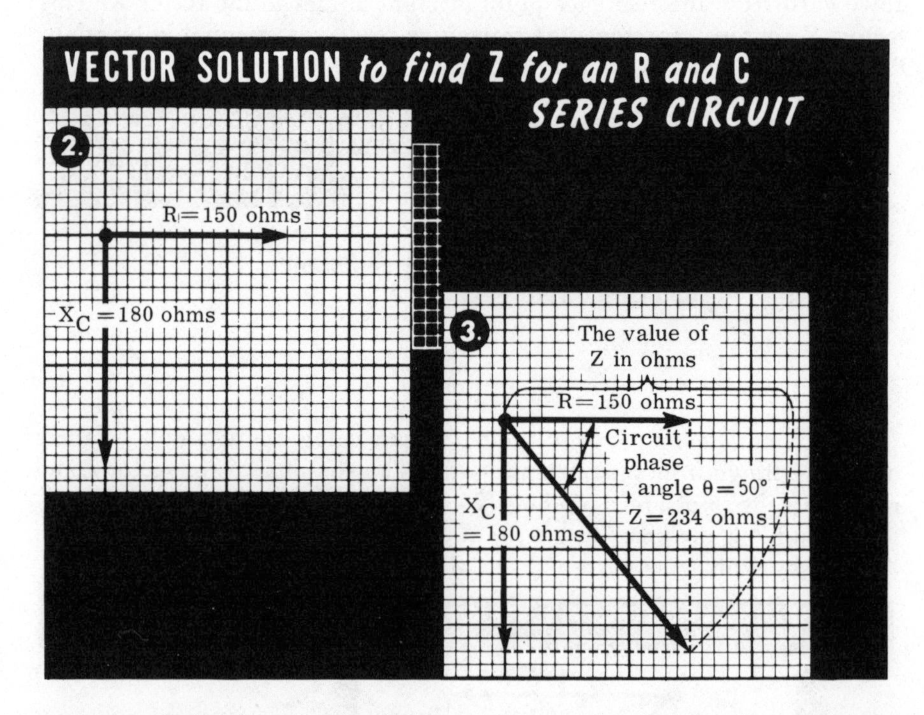

3. Using dotted lines, a parallelogram is completed and a diagonal drawn from the reference point to the intersection of the dotted lines. The length of this diagonal represents the impedance Z of 234 ohms, as measured to the same scale as that used for R and X_C. The angle between vectors R and Z is the phase angle (θ) between the circuit current and voltage.

4. Measure the angle in degrees with a protractor: θ = 50 degrees.

R and C Series Circuit Impedance (continued)

The impedance of an R and C series circuit may also easily be found by the application of the Pythagorean theorem.

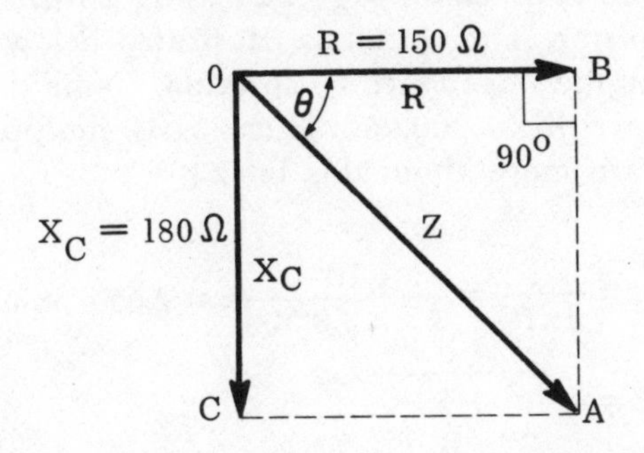

From the vector diagram above, you can see that X_C = AB and that

$$Z^2 = R^2 + X_C^2$$

$$Z = \sqrt{R^2 + X_C^2} = \sqrt{150^2 + 180^2} = \sqrt{22{,}500 + 32{,}400}$$

$$= \sqrt{54{,}900} = 234$$

$$\tan\theta = \frac{X_C}{R} = \frac{180}{150} = 1.2 \qquad \theta = \tan^{-1} 1.2 = -50°$$

A minus sign indicates that the phase angle is *below* the axis. Remember that in this case the effect of capacitive reactance (X_C) is to cause the voltage to lag the current by the phase angle (θ).

When the value of either R or X_C is unknown, but the values of the other and of Z are known, you can find the unknown value either by vector solution or by substituting known values in the formulas above.

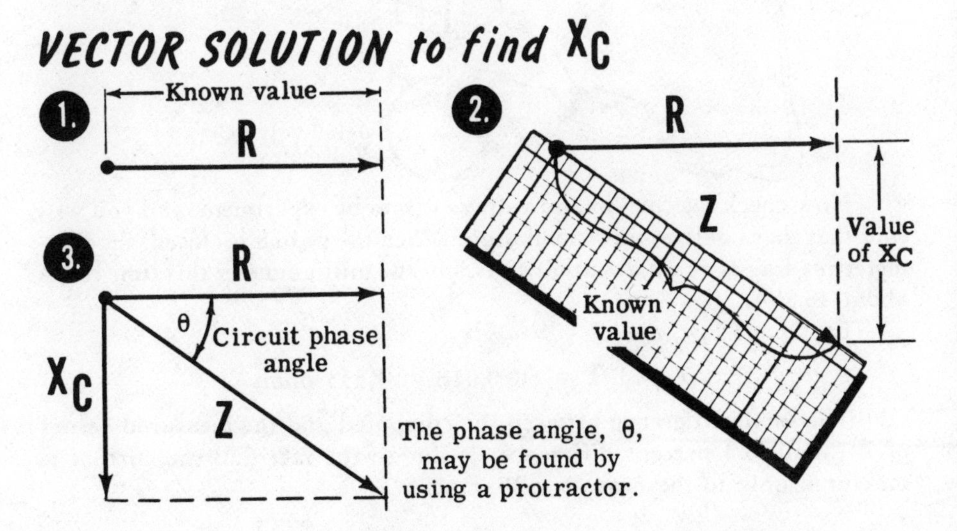

Experiment/Application—Impedance in R and C Series Circuits

You can verify that the methods of calculating impedance described on pages 4-25 and 4-26 are true for capacitive circuits by repeating the experiment/application outlined on page 4-21 using a 1-μF capacitor instead of the 5-henry inductor. The circuit is illustrated below.

First, calculate the impedance by formula. (Note that 10^6 is only a convenient way of writing 1,000,000. It means 10 multiplied by itself six times. You will learn more about this later.)

$$X_C = \frac{1}{2\pi fC} = \frac{10^6}{2\pi \times 60 \times 1} = 2{,}654 \text{ ohms}$$

$$Z = \sqrt{R^2 + X_C{}^2}$$

$$= \sqrt{2{,}000^2 + 2{,}654^2}$$

$$= \sqrt{4{,}000{,}000 + 7{,}043{,}716}$$

$$= \sqrt{11{,}043{,}716}$$

$$= 3{,}323 \text{ ohms}$$

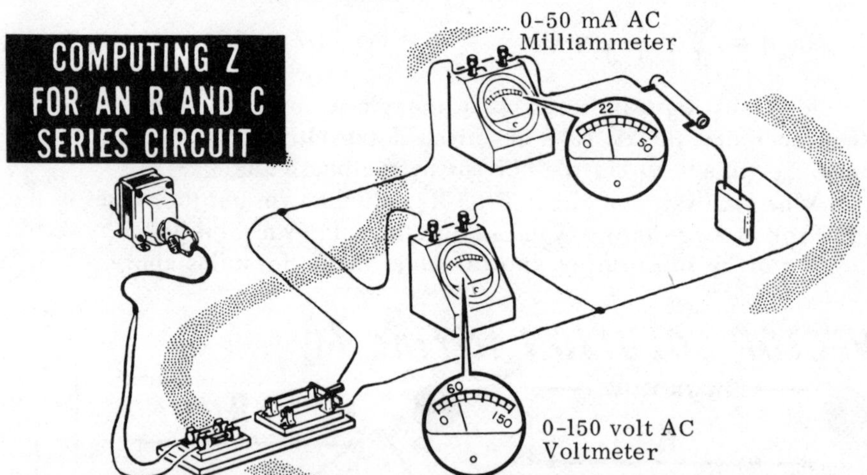

Now check the current and voltage values by experiment, and you will find that they confirm this calculation. When the switch is closed, the voltmeter reading is again about 60 volts, but the milliammeter this time reads about 18 mA.

By Ohm's law, then,

$$Z = E/I = 60/0.018 = 3{,}333 \text{ ohms}$$

The small difference between the computed and the measured values of Z (about 0.3 percent difference) is due to the fact that the current is measured only to the nearest milliampere.

R and C Series Circuit Impedance (continued)

The ratio of R to X_C determines both the amount of impedance and the phase angle in series circuits consisting only of resistance and capacitance. If the capacitive reactance is a fixed value and the resistance is varied, the impedance varies as shown below. When the resistance is near zero, the phase angle is nearly 90 degrees, and the impedance is almost entirely due to the capacitive reactance; but when R is much greater than X_C, the phase angle approaches zero degrees, and the impedance is affected more by R than by X_C.

How Impedance Varies

WHEN X_C IS A FIXED VALUE AND R IS VARIED

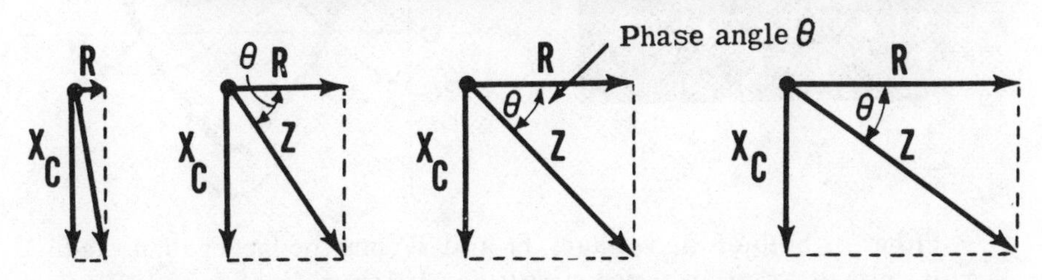

... Impedance increases and phase angle decreases as R increases

If the circuit consists of a fixed value of resistance and the capacitance is varied, the impedance varies as shown below. As the capacitive reactance is reduced toward zero, the phase angle approaches zero degrees, and the impedance is almost entirely due to the resistance. But as X_C is increased to a much greater value than R, the phase angle approaches 90 degrees, and the impedance is affected more by X_C than by R.

How Impedance Varies

WHEN R IS A FIXED VALUE AND X_C IS VARIED

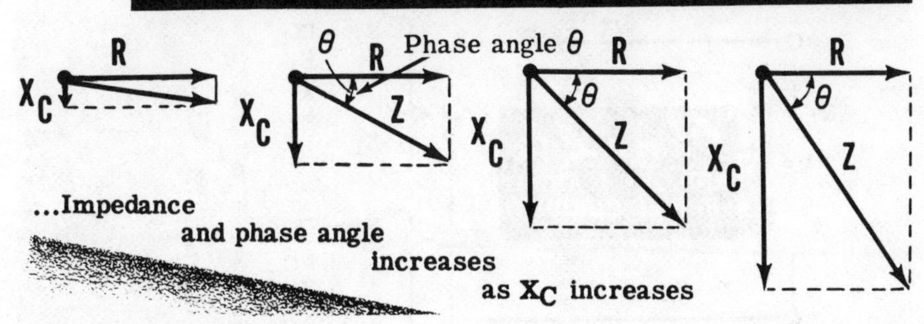

L and C Series Circuit Voltages

To find the total voltage of an L and C series circuit, you need only find the *difference* between E_L and E_C since they oppose each other directly. (Their vectors are 180 degrees apart or in exactly opposite directions.) E_L leads the circuit current by 90 degrees, while E_C lags it by 90 degrees. When the voltage waveforms are drawn, the total voltage is the arithmetic difference between the two individual values and is in phase with the larger of the two voltages—E_L or E_C. For such circuits, the total voltage can be found by subtracting the smaller voltage from the larger.

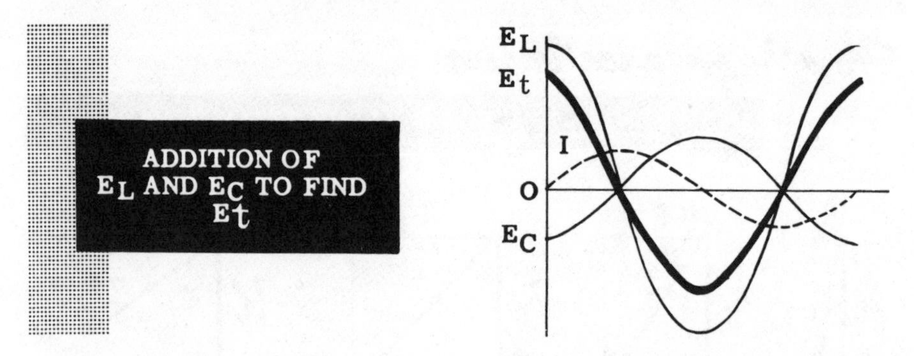

Either or both of the voltages E_L and E_C may be larger than the total circuit voltage in an ac series circuit consisting only of L and C.

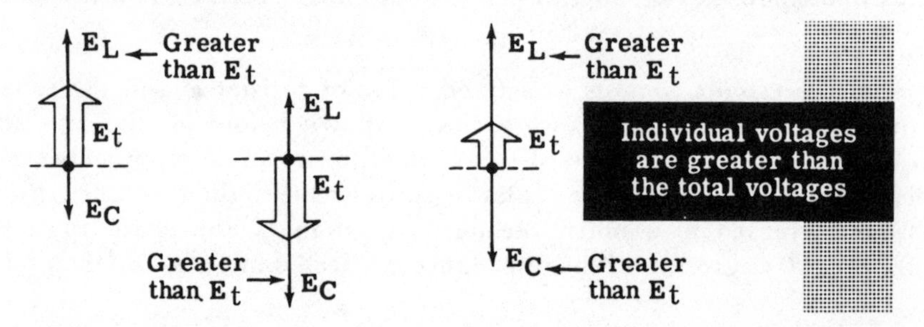

The voltage vectors and reactance vectors for the L and C circuit are similar to each other, except for the units by which they are measured. Ohm's law applies to each part, and to the total circuit outlined below.

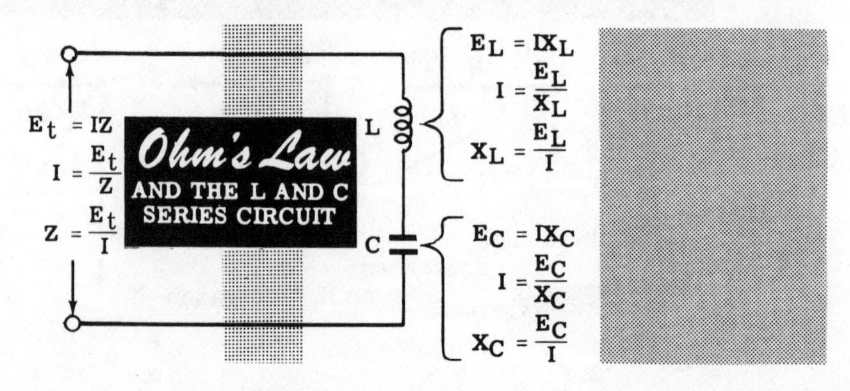

Experiment/Application—L and C Series Circuit Voltages

Suppose that you connected together a 1-μF capacitor and a 5-henry inductor to form an L and C series circuit having negligible resistance. With the power applied, if you measure individually the voltages across the inductor and the capacitor, and the total voltage across the series circuit, you will find that the voltage across the capacitor alone is about 207 volts—a value greater than the measured total voltage across the circuit. You will also find that the voltage across the inductor is about 147 volts. However, if you add the voltages vectorially (207 − 147), you will get the original supply voltage of 60 volts.

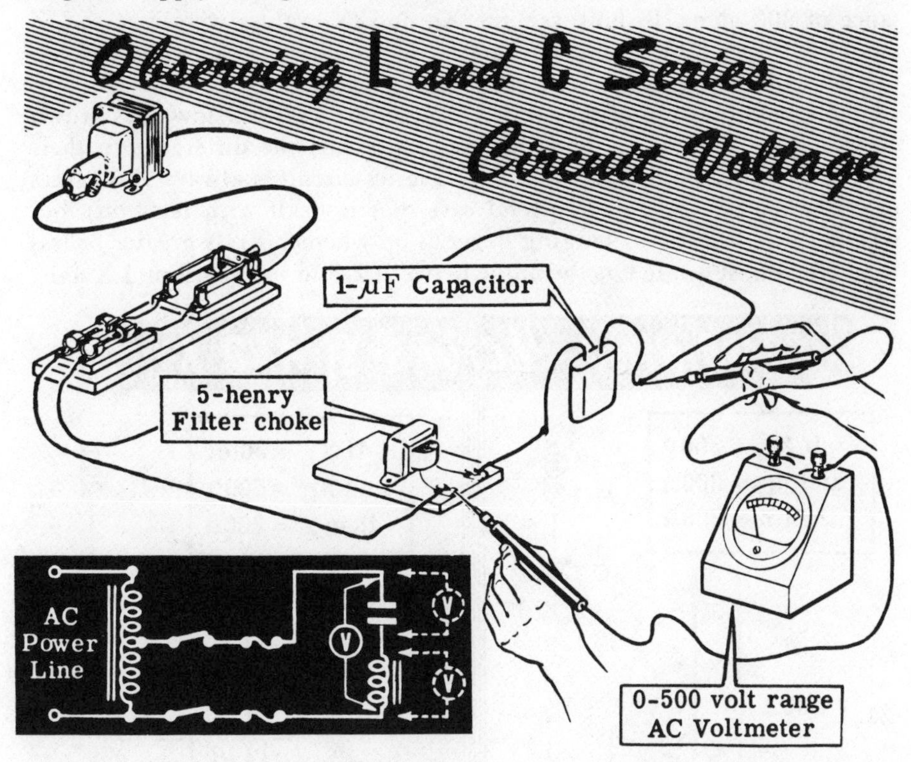

Thus, using vectors to combine the two voltages, you see that the result is approximately equal to the measured total voltage, or about 60 volts. (Although it is considered negligible, the resistance of the inductor wire does, in practice, cause a slight difference between the computed and the actual results.)

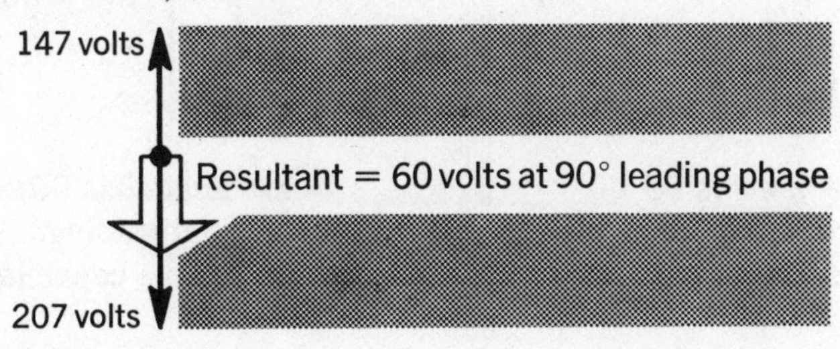

L and C Series Circuit Impedance

In ac series circuits consisting of inductance and capacitance, with only negligible resistance, the impedance is due only to inductive and capacitive reactance. Since inductive and capacitive reactances act in *opposite* directions, the total effect of the two is equal to the *difference* between them. For such circuits, Z can be found by subtracting the smaller value from the larger. The circuit will then act as an inductive or a capacitive reactance (depending on which is larger) that has an impedance equal to Z. For example, if $X_L = 500$ ohms and $X_C = 300$ ohms, the impedance Z is 200 ohms, and the circuit acts as an *inductance* having an inductive reactance of 200 ohms. If, however, the X_L and X_C values were reversed, Z would still equal 200 ohms, but the circuit would act as a *capacitance* having a capacitive reactance of 200 ohms.

The relationships of the above examples are shown below. Z is drawn on the same axis as X_L and X_C and represents the difference in their values. The phase angle of the L and C series circuit is always 90 degrees except when $X_L = X_C$ (a special case that is dealt with later on), but whether it is leading or lagging depends on whether X_L is greater or less than X_C. Phase angle θ is the angle between Z and the horizontal X axis.

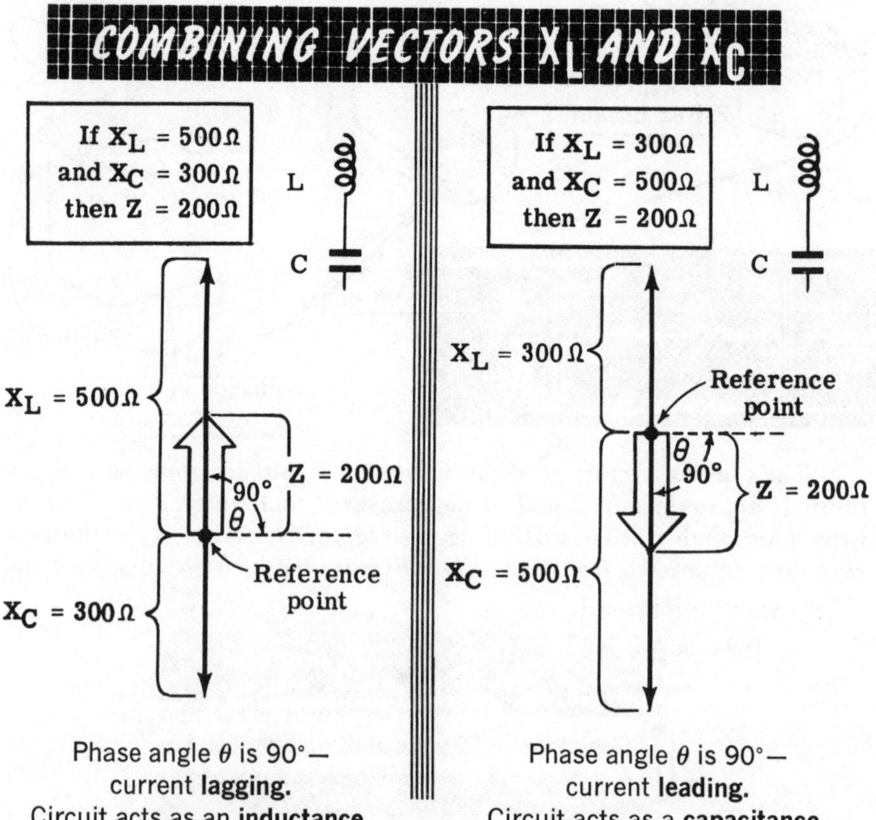

COMBINING VECTORS X_L AND X_C

If $X_L = 500\Omega$
and $X_C = 300\Omega$
then $Z = 200\Omega$

L
C

$X_L = 500\Omega$
$Z = 200\Omega$
90°
θ
Reference point
$X_C = 300\Omega$

Phase angle θ is 90°—
current **lagging.**
Circuit acts as an **inductance.**

If $X_L = 300\Omega$
and $X_C = 500\Omega$
then $Z = 200\Omega$

L
C

$X_L = 300\Omega$
Reference point
θ
90°
$Z = 200\Omega$
$X_C = 500\Omega$

Phase angle θ is 90°—
current **leading.**
Circuit acts as a **capacitance.**

Experiment/Application—L and C Series Circuit Impedance

To see how inductive and capacitive reactance oppose each other, connect a 1-μF capacitor across the secondary of a 120-volt to 60-volt step-down transformer. (The circuit is illustrated below.) With the switch closed, you will see that the voltage is about 60 volts and that the current is about 22.6 mA. The Ohm's law value of the impedance is then approximately 2,655 ohms ($60 \div 0.0226 \approx 2.655$). Since only capacitance is used in the circuit, the value of Z is equal to X_C.

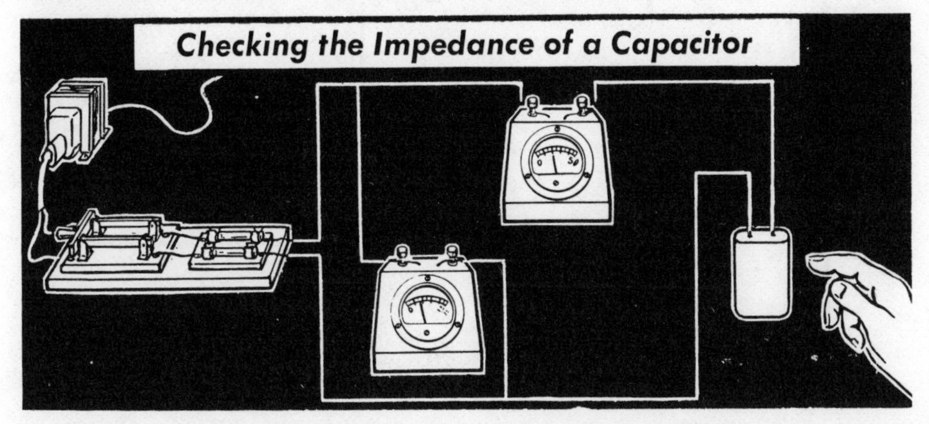

If you now replace the capacitor with a 5-henry inductor in the same circuit, you will find the circuit current to be about 31.8 mA, and the current impedance is then $60 \div 0.0318$, or approximately 1,887 ohms.

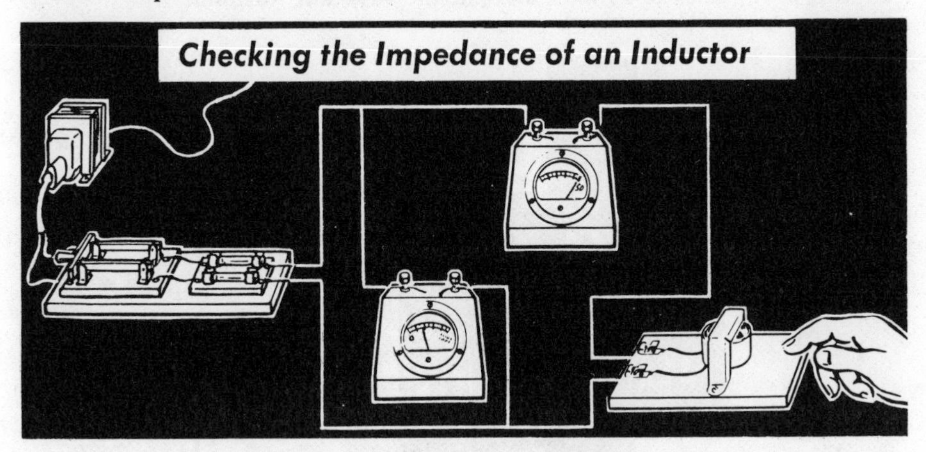

If you now redo the experiment on page 4-31 with an ammeter in the circuit where both the 5-henry inductance and the 1-μF capacitor are in series, you will find that the circuit current is 0.078 amp. The net circuit impedance is about 770 ohms ($Z = E/I = 60/0.078 = 770$). You will note that 770 ohms is approximately the difference between 2,655 ohms and 1,887 ohms. Thus, it is clear that the net circuit impedance is the difference between the inductive and capacitive reactances.

R, L, and C Series Circuit Voltages

To combine the three voltages of an R, L, and C series circuit by means of vectors, two steps are required:

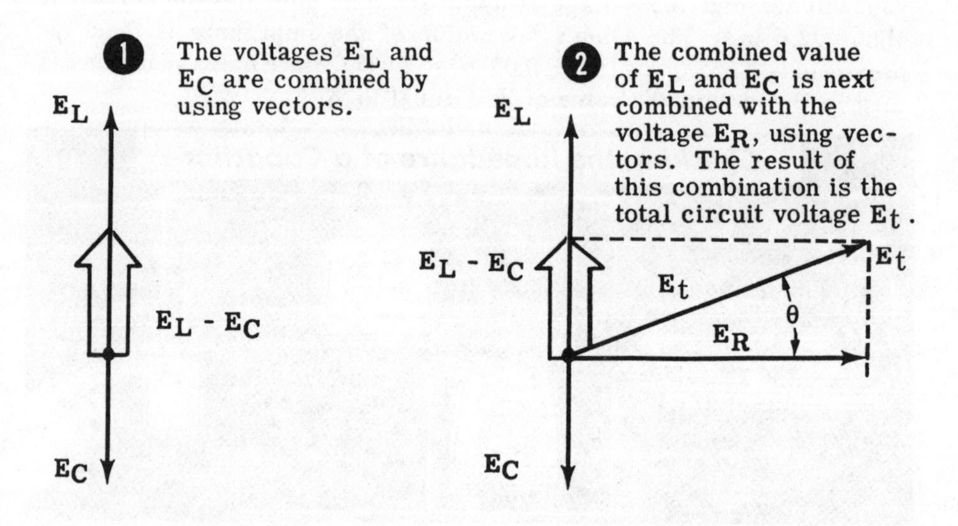

You can apply Ohm's law for any part of the circuit by using X_L, X_C, or R across inductors and capacitors or resistors, respectively. For the total circuit, Z replaces R as it is used in the original formula.

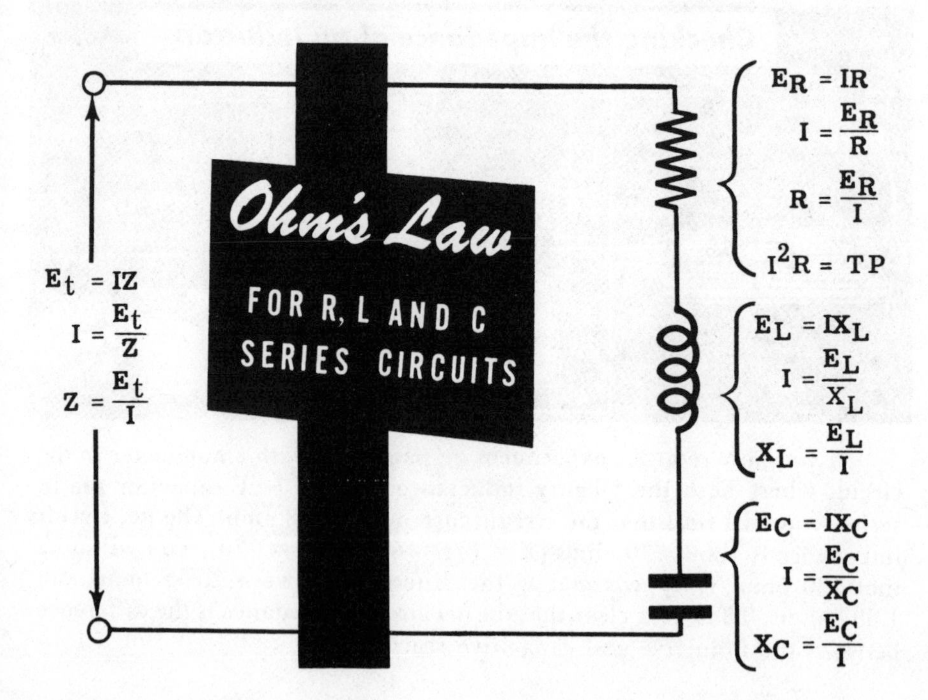

R, L, and C Series Circuit Impedance—Graphic Solution

The impedance of a series circuit consisting of resistance, capacitance, and inductance in series depends on R, X_L, and X_C. If the values of all three factors are known, impedance Z may be found as follows:

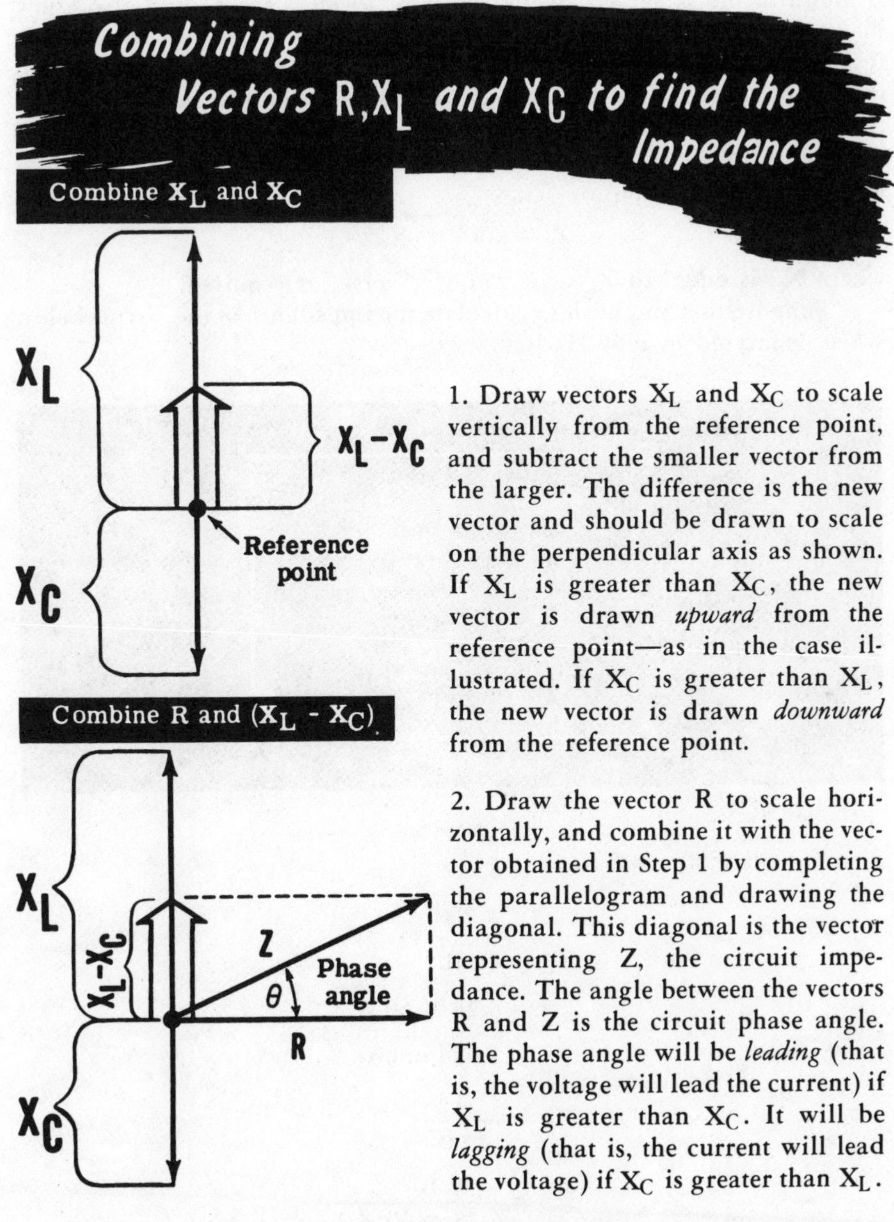

1. Draw vectors X_L and X_C to scale vertically from the reference point, and subtract the smaller vector from the larger. The difference is the new vector and should be drawn to scale on the perpendicular axis as shown. If X_L is greater than X_C, the new vector is drawn *upward* from the reference point—as in the case illustrated. If X_C is greater than X_L, the new vector is drawn *downward* from the reference point.

2. Draw the vector R to scale horizontally, and combine it with the vector obtained in Step 1 by completing the parallelogram and drawing the diagonal. This diagonal is the vector representing Z, the circuit impedance. The angle between the vectors R and Z is the circuit phase angle. The phase angle will be *leading* (that is, the voltage will lead the current) if X_L is greater than X_C. It will be *lagging* (that is, the current will lead the voltage) if X_C is greater than X_L.

You can always find the impedance of any circuit by applying Ohm's law for ac circuits, after measuring the circuit current and voltage.

R, L, and C Series Circuit Impedance—by Calculation

The impedance of a circuit which contains R, L, and C components may also be calculated by using a variation of the impedance formula, $Z = \sqrt{R^2 + X^2}$. You have learned that it makes no difference whether the reactive component, X, is inductive or capacitive in nature; the impedance is found in the same way, using the same formula for Z. You also know that, when both inductive and capacitive reactances are present in a circuit, it is only necessary to subtract the smaller amount of reactance (either inductive or capacitive, as the case may be) from the larger amount and then to draw in the resultant diagonal vector Z. In calculating the value for impedance in a circuit containing both inductive and capacitive components, use the formula,

$$Z = \sqrt{R^2 + X_e^2}$$

where X_e is equal to $X_L - X_C$ or vice versa, as required.

Suppose that you wish to calculate the impedance of the circuit below when connected to a 60-Hz line:

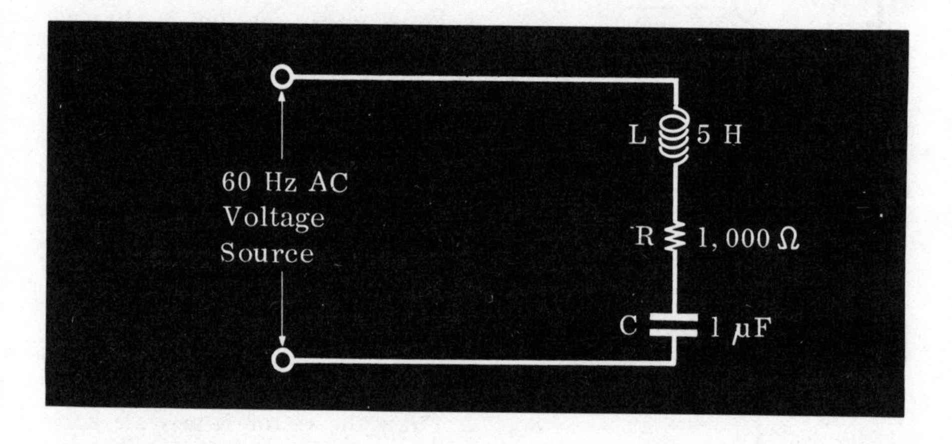

First calculate X_L and X_C:

$$\begin{aligned} X_L &= 2\pi fL \\ &= 6.28 \times 60 \times 5 \\ &= 1{,}884 \text{ ohms} \end{aligned}$$

$$\begin{aligned} X_C &= \frac{1}{2\pi fC} \\ &= \frac{10^6}{6.28 \times 60 \times 1} \\ &= 2{,}654 \text{ ohms} \end{aligned}$$

R, L, and C Series Circuit Impedance—by Calculation (continued)

(Note that 10^6 is only a convenient way of writing 1,000,000. It means 10 multiplied by itself six times. You will learn more about this later.)

$$
\begin{aligned}
Z &= \sqrt{R^2 + (X_L - X_C)^2} \\
&= \sqrt{1{,}000^2 + (1{,}884 - 2{,}654)^2} \\
&= \sqrt{1{,}000^2 + (-770)^2} \\
&= \sqrt{10^6 + 592{,}900} \\
&= \sqrt{1{,}592{,}900} \\
&= 1{,}262 \text{ ohms}
\end{aligned}
$$

As X_C is greater than X_L, the current leads the total voltage across the circuit.

The phase angle can be calculated as before from

$$\tan\theta = \frac{X}{R}$$

where X is the net circuit reactance ($X_L - X_C$). Since $X_L - X_C = 1{,}885 - 2{,}654 = -770$,

$$\theta = \tan^{-1}\frac{X}{R} = \tan^{-1}\frac{-770}{1{,}000} = -37.6^{\circ}$$

and since X is negative, the phase angle is negative, indicating that the net circuit reactance is capacitive—that is, the voltage lags the current.

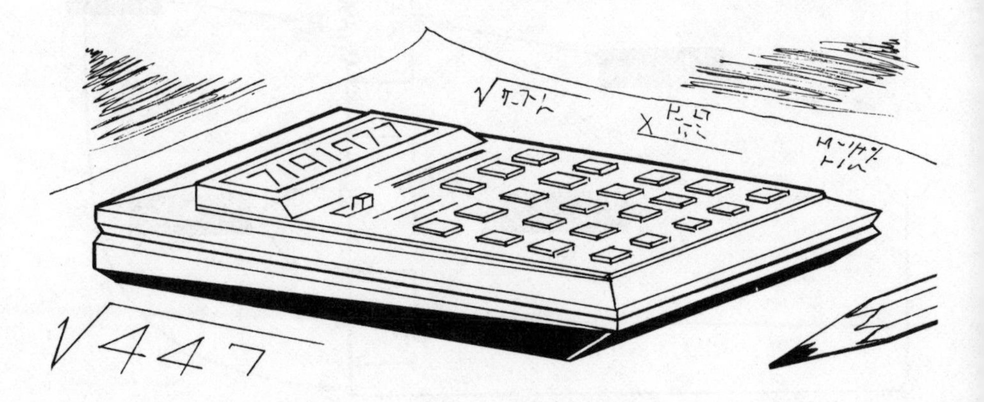

Series Circuit Resonance

As you may have suspected, in any series circuit containing both L and C, the current is greatest when the inductive reactance X_L equals the capacitive reactance X_C, since under those conditions the impedance is equal to R. Whenever X_L and X_C are unequal, the impedance Z is the vector combination of R and the difference between X_L and X_C. This vector is always greater than R, as shown below. When X_L and X_C are equal, Z is equal to R and is at its minimum value, allowing the greatest amount of current to flow. As you can see, if R is small, very large amounts of current can flow.

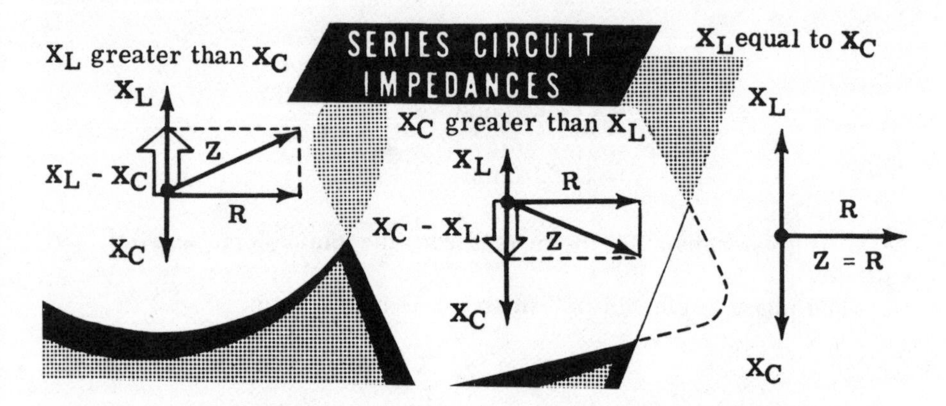

When X_L and X_C are equal, the voltages across them, E_L and E_C, are also equal, but of opposite phase, and the circuit is said to be *at resonance.* Such a circuit is called a *series resonant circuit.* (Note that both E_L and E_C may be much greater than E_t, which equals E_R at resonance.)

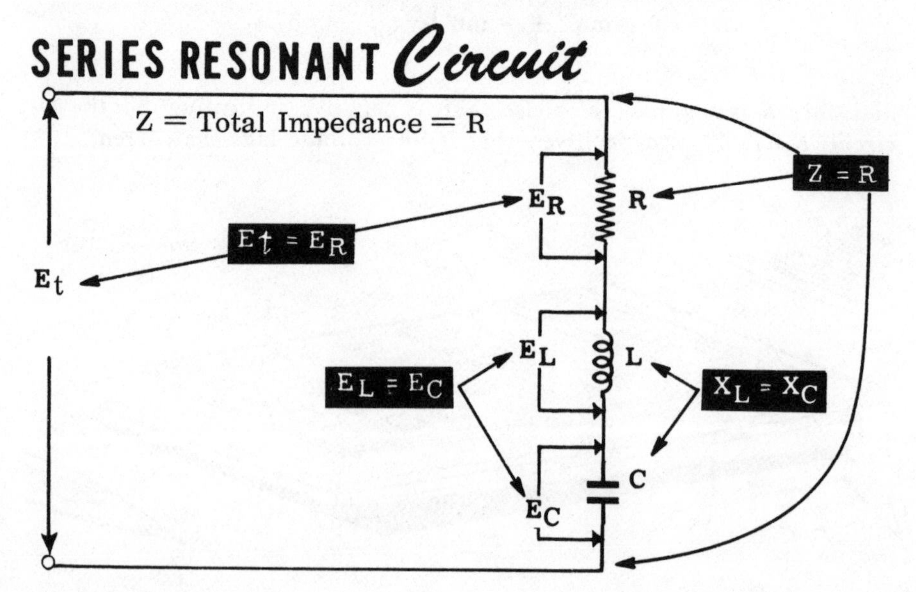

Series Circuit Resonance—Resonant Frequency

Suppose now that an ac voltage of variable frequency is applied to a circuit consisting of an inductor and a capacitor in series. The inductor is represented in the circuit diagram below by a pure inductance (L) in series with its internal resistance (R).

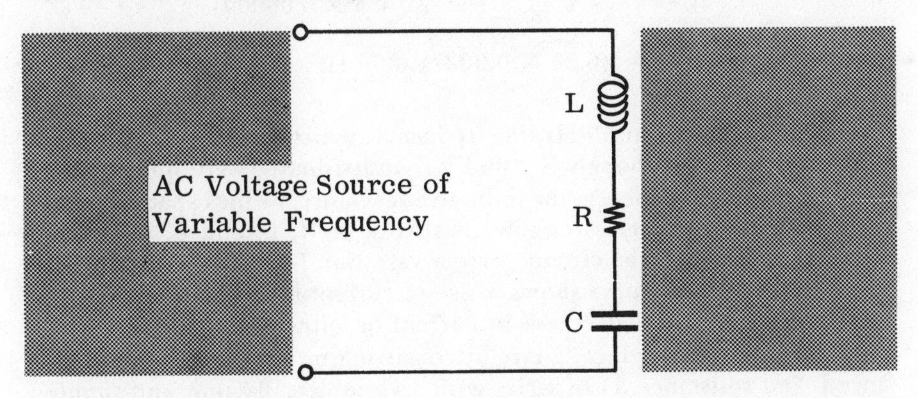

As the frequency of the applied voltage rises, the capacitive reactance X_C decreases, but the inductive reactance X_L increases. The graphs which follow show how X_C and X_L vary with changes in the frequency.

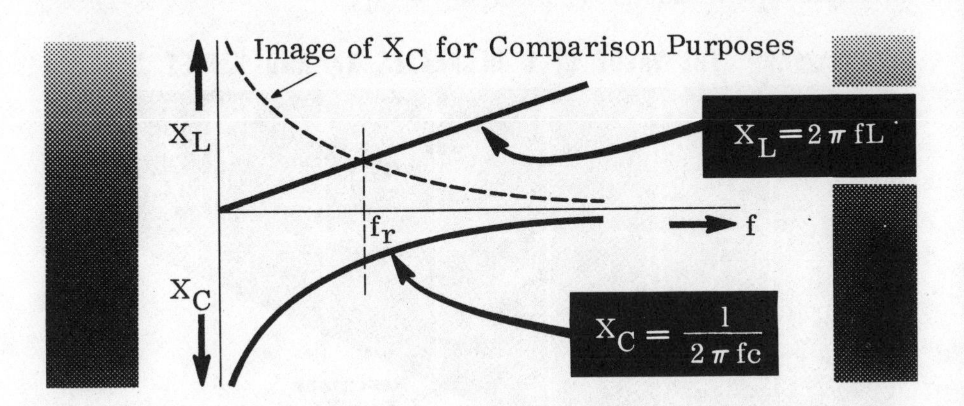

You can see that at only one frequency does X_C equal X_L. This is called the *resonant frequency, f_r*. The frequency at which a particular circuit will resonate can be calculated as follows:

$$X_C = X_L$$
$$2\pi f_r L = 1/2\pi f_r C$$
$$f_r^{\,2} = 1/2^2\pi^2 LC$$
$$f_r = 1/2\pi\sqrt{LC}$$

As you can see, the resonant frequency is *not* affected by R.

Series Circuit Resonance—Resonant Frequency (continued)

For our circuit, with a 5-henry inductor and a 1-μF capacitor, we can calculate the resonant frequency as

$$f_r = 1/2\pi\sqrt{LC} = 1/2\pi\sqrt{5 \times 0.000001}$$

$$= 1 / (6.28 \times 0.0022) = 71 \text{ Hz}$$

This is not far from our 60-Hz line frequency, which accounts for the large circuit current even though X_L and X_C individually were quite high.

If either the frequency, the inductive reactance, or the capacitive reactance is varied in a series circuit consisting of R, L, and C, with other values kept constant, the circuit current variation forms a curve called the *resonance curve.* This curve shows a rise in current to a maximum value at exact resonance and a decrease in current on either side of resonance.

Consider, for instance, a circuit consisting of an inductor (of inductance L and resistance R) in series with a variable capacitor, and suppose that this circuit is connected across a suitable ac voltage source whose frequency is fixed. As the capacitor is varied from its minimum capacitance to its maximum capacitance, the impedance of the circuit will vary as illustrated below. You will notice that the current reaches its maximum, and the impedance, its minimum, when $X_C = X_L$.

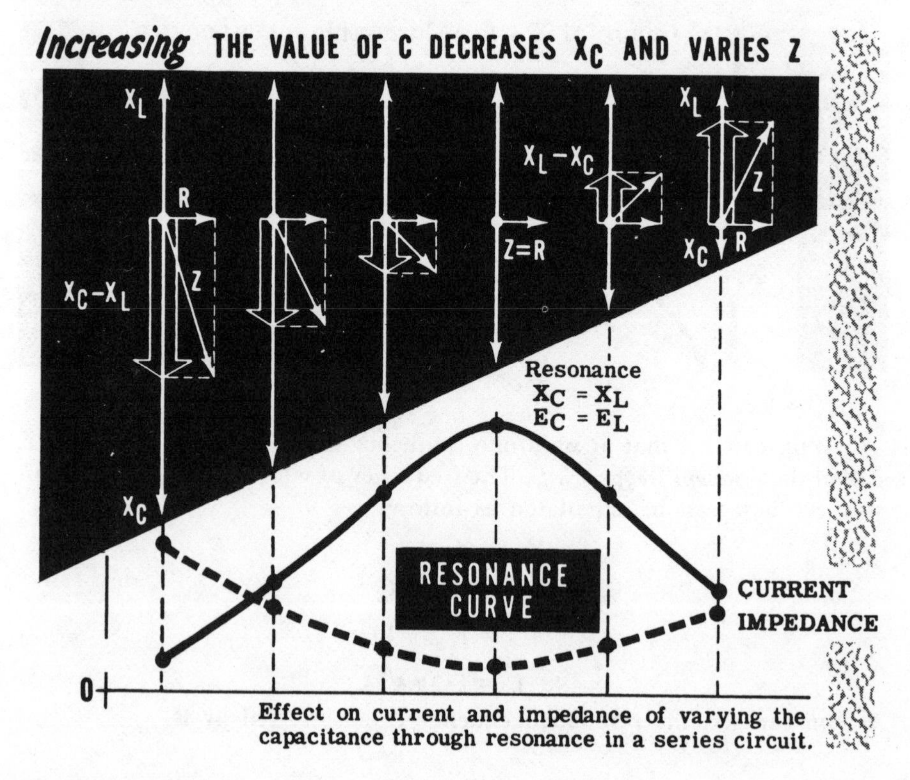

Effect on current and impedance of varying the capacitance through resonance in a series circuit.

Series Circuit Resonance—Resonant Frequency (continued)

Similar curves result if capacitance and frequency are held constant while the inductance is varied, and also if the inductance and capacitance are held constant while the frequency is varied.

Variation of the voltage input frequency to an R, L, and C circuit (over a suitable range) results in an output current curve which is similar to the resonance curve on the previous page. When operated below the resonant frequency, the current is low and the circuit impedance is high. Above resonance, the same condition occurs. At the resonant frequency, the impedance curve has its minimum value, equal to that of the circuit resistance; the current curve must, therefore, be at its maximum value (given, of course, by the equation $I = E/R$). The graphs of current and impedance below are typical of an R, L, C series circuit when only the applied voltage frequency is varied.

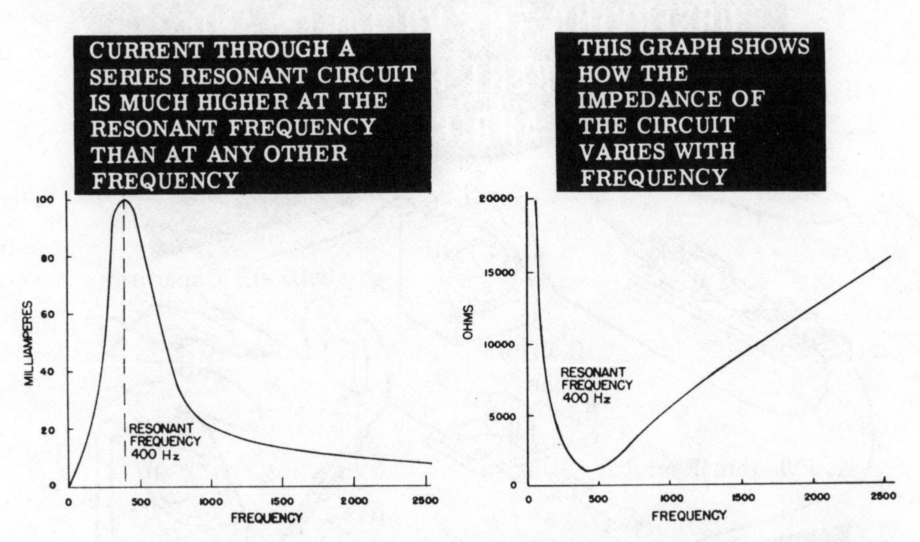

Note that below the resonant frequency the circuit acts as a capacitance because X_C is greater than X_L, and the current leads the applied voltage. Above the resonant frequency, the circuit acts as an inductance because X_L is greater than X_C, and the voltage leads the current. The variation of phase angle is illustrated below.

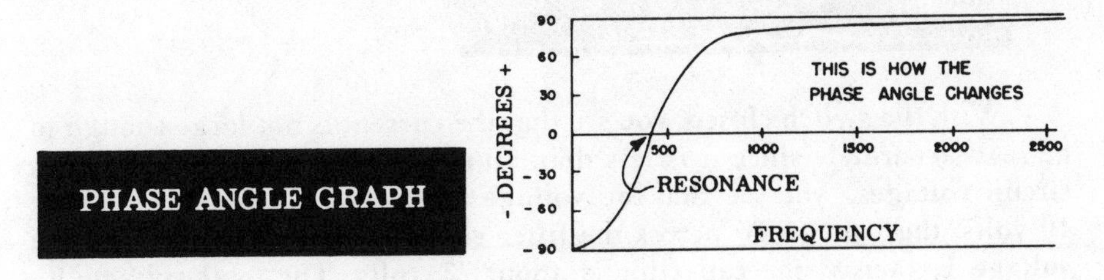

At resonance, the phase angle is zero.

Experiment/Application—Series Resonance

As you know from the earlier experiment/application, the series L-C circuit with a 5-henry inductor and 1-μF capacitor is resonant at about 71 Hz. To make it resonant at 60 Hz, you could increase either L or C. In this experiment/application, you will explore the effects of series resonance on the circuit impedance.

To demonstrate series resonance, suppose that you replaced the 1-μF capacitor with a 0.25-μF capacitor and inserted a 1,500-ohm, 10-watt resistor in series with the 5-henry filter choke and the capacitor. This forms an R, L, and C series circuit, in which C will be varied to show the effect of resonance on circuit voltage and current. A 0–50 mA ac milliammeter is connected in series with the circuit to measure the circuit current. A 0–250 volt ac voltmeter measures circuit voltages.

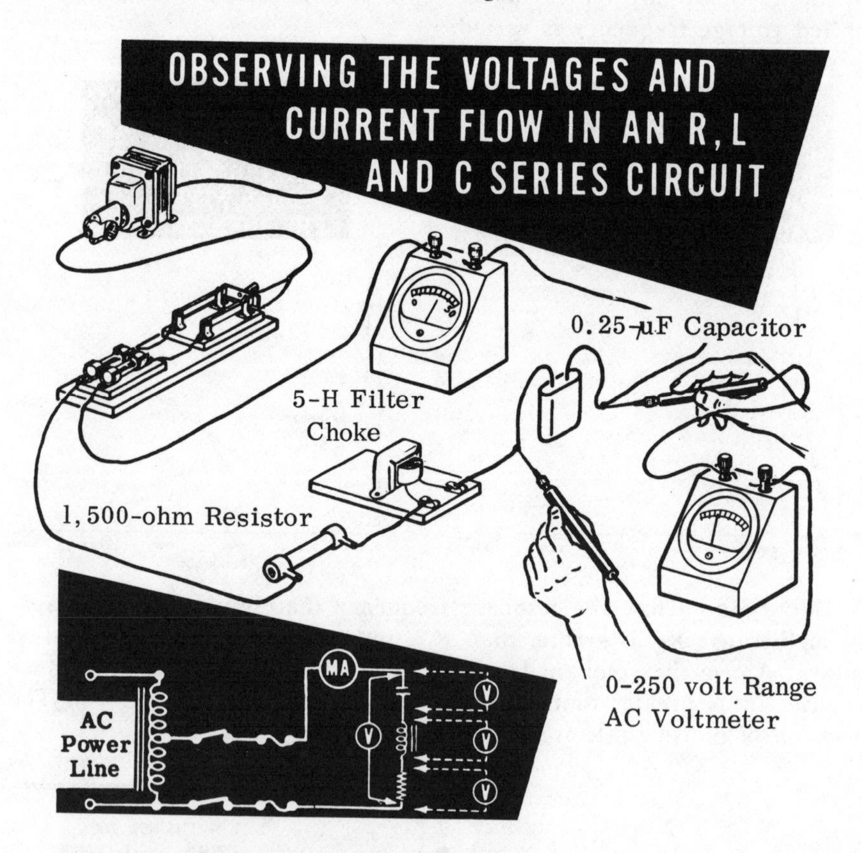

With the switch closed, you see that the current is not large enough to be read accurately since it is less than 10 mA. If you measure the various circuit voltages, you see that the voltage E_R across the resistor is about 10 volts, the voltage E_L across the filter choke is about 13 volts, and the voltage E_C across the capacitor is about 72 volts. The total voltage E_t across the entire circuit is approximately 60 volts.

Experiment/Application—Series Resonance (continued)

By using various parallel combinations of 0.25-μF, 0.5-μF, 1-μF, and 2-μF capacitors, you can vary the circuit capacitance from 0.25 μF to 3.75 μF in steps of 0.25 μF. It is important to remove the circuit power and discharge all capacitors used before removing or adding capacitors to the circuit. You will notice that as the capacitance is increased, the current rises to a maximum value at the point of resonance, then decreases as the capacitance is increased further.

OBSERVING THE CIRCUIT VOLTAGES AND CURRENT CHANGE AS THE CAPACITANCE VALUE IS CHANGED

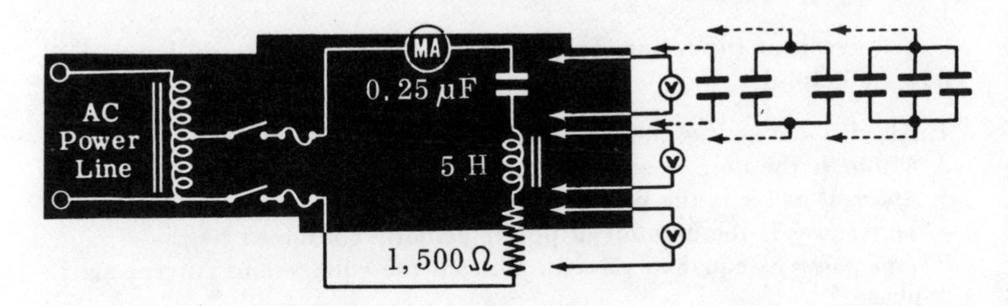

Except for the total circuit voltage, E_t, which is constant at 60 volts, the measured circuit voltages vary as the capacitance is changed. The voltage E_R across the resistor changes in the same manner as the circuit current. For capacitance values less than the resonance value, E_C is greater than E_L. At resonance, E_L equals E_C. As the capacitance is increased beyond the resonance point, E_L is greater than E_C when the circuit capacitance is greater than the value required for resonance.

You could, of course, have calculated the correct value from the resonance formula that you learned earlier, or, knowing that at resonance $X_L = X_C$, you can calculate as follows:

$$X_L = 2\pi fL$$

$$X_C = 1/2\pi fC$$

$$2\pi fL = 1/2\pi fC$$

$$2 \times 3.14 \times 60 \times 5 = 1/2 \times 3.14 \times 60 \times C$$

$$C = 1/(2 \times 3.14 \times 60 \times 5) \times (2 \times 3.14 \times 60)$$

$$C = 1/709{,}891 = 0.00000141 = 1.41\ \mu F$$

As you saw from the experiment, the capacitance value that gave the highest current in the circuit and the maximum voltage across the resistance was equal to 1.5 μF (close to 1.4).

Power Factor

The concept of power, particularly with regard to the calculation of power factors in ac circuits, will become a very important consideration as the circuits you work with are made more complex. In dc circuits, the expended power may be determined by multiplying the voltage by the current. A similar relationship exists for finding the amount of expended power in ac circuits, but certain other factors must be considered. For example, you know that pure inductive or capacitive ac circuits can draw current although the power consumed is zero. In pure resistive ac circuits, the power is equal to the product of the voltage and current *only* when the E and I are in phase or at resonance. If the voltage and current are not in phase, the power used by the circuit will be less than the product of E and I.

A recap of the following principles, which you already know, will now be helpful:

1. *Power* is defined as the rate of doing work.
2. A *watt* is the unit of electrical power.
3. *Apparent power* is the product of volts and amperes in an ac circuit.
4. *True power* is the amount of power actually consumed by the circuit.
5. *True power* is equal to *apparent power* if the voltage and current are in phase.
6. *True power* is equal to zero if the voltage and current are out of phase by 90 degrees.

Now consider again the vector diagram for the voltages in an R and L ac series circuit. E_L is the voltage across the inductance, E_R is the voltage across the resistance, and E_t is the total applied voltage.

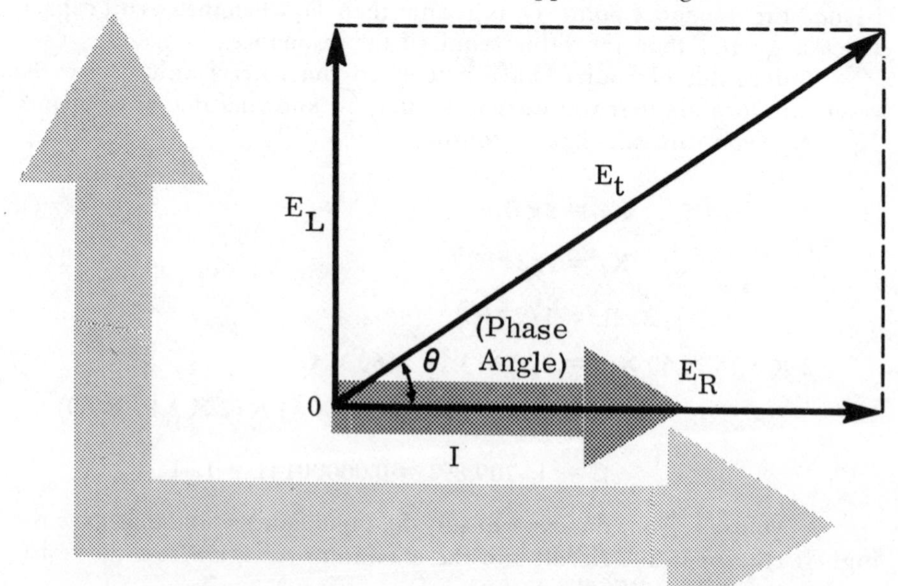

Power Factor (continued)

You can see that E_R is the component of E_t which is in phase with the current I, and E_L is the component of E_t which is 90 degrees out of phase with I.

From the definitions of true power and apparent power repeated above, we can see that:

$$\text{True power} = E_R I$$
$$\text{Apparent power} = E_t I$$

We define the "power factor" as the ratio of true to apparent power:

$$\text{Power factor} = \frac{\text{True power}}{\text{Apparent power}} = \frac{E_R I}{E_t I}$$

If we cancel the I in the numerator and denominator, we get:

$$\text{Power factor} = \frac{E_R}{I} \times \frac{I}{E_t} = \frac{E_R}{E_t} = \frac{R}{Z}$$

If you check back on your study of the right triangle, you will see that the ratio E_R/E_t is the cosine of angle θ, which is the phase angle. Thus, power factor = $\cos\theta$, and

$$\text{True power (P)} = \text{apparent power} \times \text{power factor}$$
$$P = E_t I \cos\theta$$

If you again compare the vector diagram of impedance, reactance, and resistance on page 4-13 with the right triangle on page 4-19, you will see that $\cos\theta$ equals R/Z. You will find the quantity power factor expressed as a percentage calculated as $(R/Z) \times 100$.

POWER FACTOR =

$$\frac{\textbf{\textit{TRUE} Power}}{\textbf{\textit{APPARENT} Power}}$$

Power

Suppose that you wish to find the amount of power expended (true power) in a circuit where the impedance is 5 ohms, the resistance is 3 ohms, and the inductive reactance is 4 ohms. The voltage is 10 volts ac, and the current is 2 amperes.

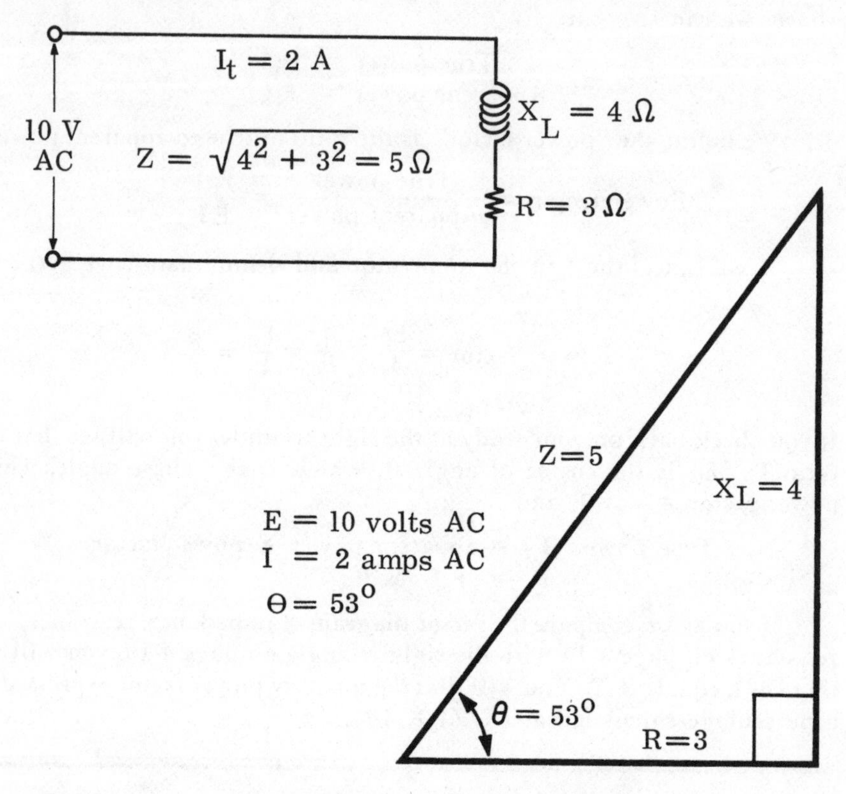

The apparent power is given by E × I, or 20 watts.

The formula for true power is $P = EI \cos \theta$. With reference to the impedance triangle, $\cos \theta$ is equal to the ratio of R divided by Z, or

$$\cos \theta = \frac{R}{Z} = \frac{3}{5} = 0.60 \text{ (60\% power factor)}$$

By looking up cos 53° in the back of this volume, we find it equal to 0.6 also. Substituting in the formula for true power,

$$\begin{aligned} \text{True power} &= EI \cos \theta \\ &= 10 \times 2 \times 0.6 \\ &= 12 \text{ watts} \end{aligned}$$

As you can see, the line current is 2 amperes but the power is only 12 watts. The line current would be only 1.2 ampere if the voltage and current were in phase. As you will learn later, this is a very important point since the power line has to carry the 2 amperes and any resistive loss is determined by the line resistance and the 2 amperes, not the 1.2 ampere.

Experiment/Application—Power and Power Factor

To demonstrate the difference between true and apparent power, you could use the series R, L, and C circuit with a wattmeter that you know measures true power. For this experiment/application, you will want to reduce the resistance to 500 ohms.

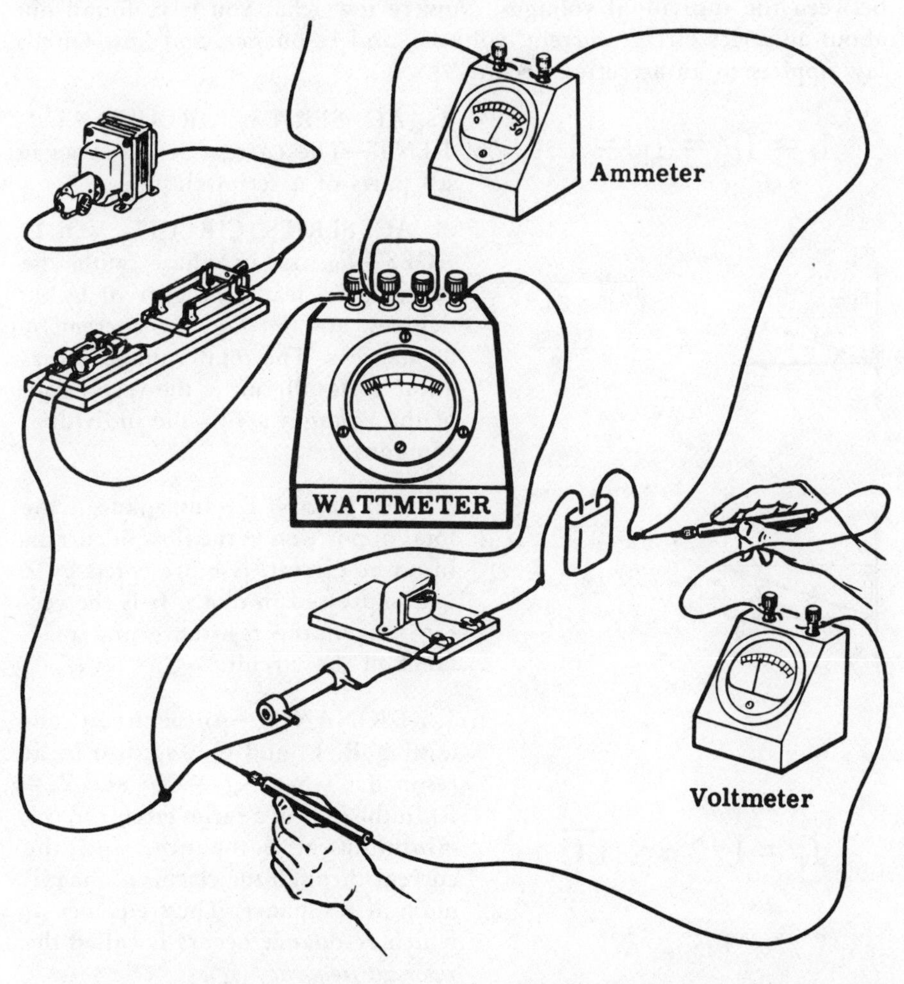

Vary the capacitance as before. You will find that the wattmeter (measuring true power) will read the true power as being equal to the current squared times the resistance (I^2R) whereas the apparent power, calculated as E × I, will not be the same. If you calculate the power factor as true power divided by apparent power for each capacitor value, you will find that they will be equal at resonance (where there is no apparent reactance in the circuit). Thus, the power factor at resonance is unity (or 100 percent). You will see this important fact made use of later, where inductance and capacitance are added to power transmission lines to correct the power factor to unity so as to obtain the most efficient power transfer.

Review of Current, Voltage, Impedance, Resonance, Power, Power Factor

You have found that the rules for ac series circuit voltage and current are the same as those for dc series circuits, except that the various circuit voltages must be *added by means of vectors* because of the phase difference between the individual voltages. Now review what you have found out about ac series circuit current, voltages, and resonance, and how Ohm's law applies to an ac series circuit.

$$I_t = I_C = I_L = I_R$$

1. AC SERIES CIRCUIT CURRENT—The current is the same in all parts of a series circuit.

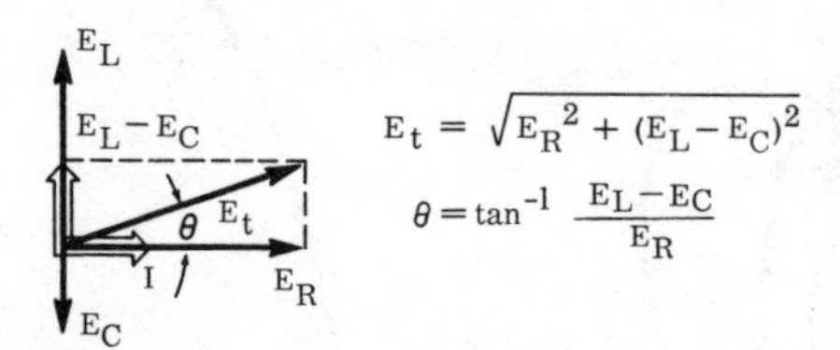

2. AC SERIES CIRCUIT VOLTAGES—E_R is in phase with the current, E_L leads the current by 90 degrees, and E_C lags the current by 90 degrees. The total voltage across an ac series circuit is the vector sum of the voltages across the individual components.

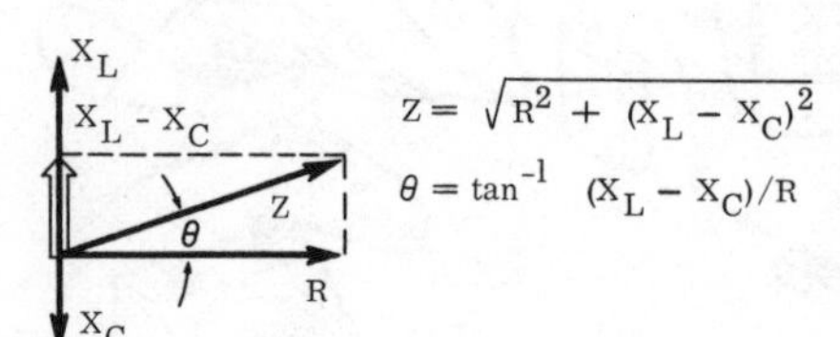

3. IMPEDANCE—Impedance, the total opposition to the flow of current in an ac circuit, is represented by Z and expressed in ohms. It is the vector sum of the resistance and reactance of the circuit.

$$f_r = 1/2\pi\sqrt{LC}$$

4. RESONANCE—An ac circuit containing R, L, and C is said to be at resonance when $X_L = X_C$ and $Z = R$. In the case of a series circuit, Z is a minimum at resonance, and the current through the circuit is a maximum at resonance. The frequency at which resonance occurs is called the *resonant frequency* (f_r).

$$\text{True power} = I^2R = EI\cos\theta$$
$$\text{Apparent power} = EI$$
$$\text{Power factor} = \frac{\text{True power}}{\text{Apparent power}}$$
$$\text{Power factor} = \cos\theta = R/Z$$

5. POWER/POWER FACTOR—The power in ac circuits containing reactances does not usually equal voltage times current (E × I) as in dc circuits. This value is called the *apparent power. True power* must also take phase into account. The ratio of true to apparent power is called the *power factor.*

Self-Test—Review Questions

1. Draw a vector diagram to represent an R of 1000 ohms, an X_L for a 5-henry inductor, and an X_C for a 2-μF capacitor at 60 Hz.
2. For the circuits shown below (A through F), draw a vector diagram solution to find Z and θ (the frequency is 60 Hz).

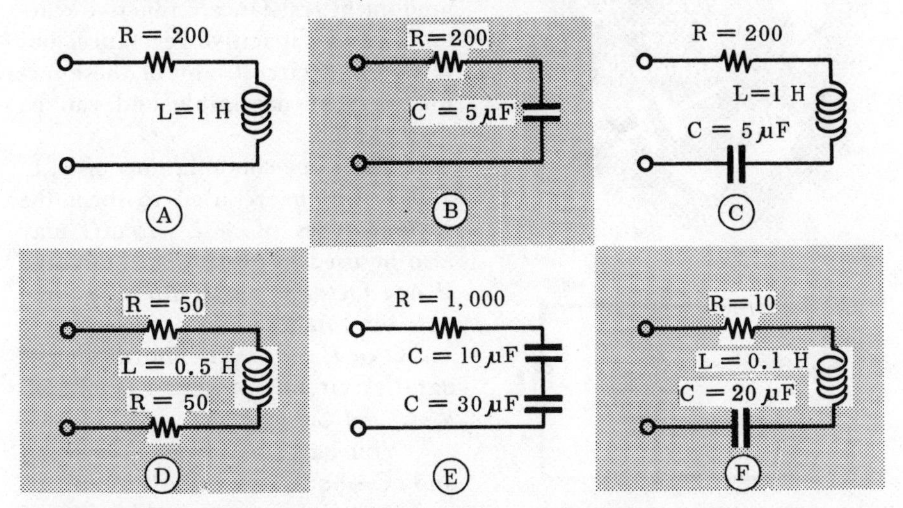

3. Solve the circuits above by calculation and compare your answers.
4. If the input voltage is 120 volts, 60 Hz, calculate the circuit currents and the voltage across each component in Question 2.
5. For each circuit in Question 2, calculate the true power, apparent power, and power factor.
6. Calculate the resonant frequencies for circuits (c) and (f). How much current will flow at resonance? What is the phase? What capacitance must be added to circuit (c) for it to be resonant at 60 Hz? What inductance must be added to circuit (c) for it to be resonant at 60 Hz?

Learning Objectives—Next Section

Overview—Now that you have learned about ac series circuits, you will learn in the next section about ac parallel circuits. They are most important since electric circuits are often made up of many devices connected in parallel across the ac line.

AC Parallel Circuit Combinations

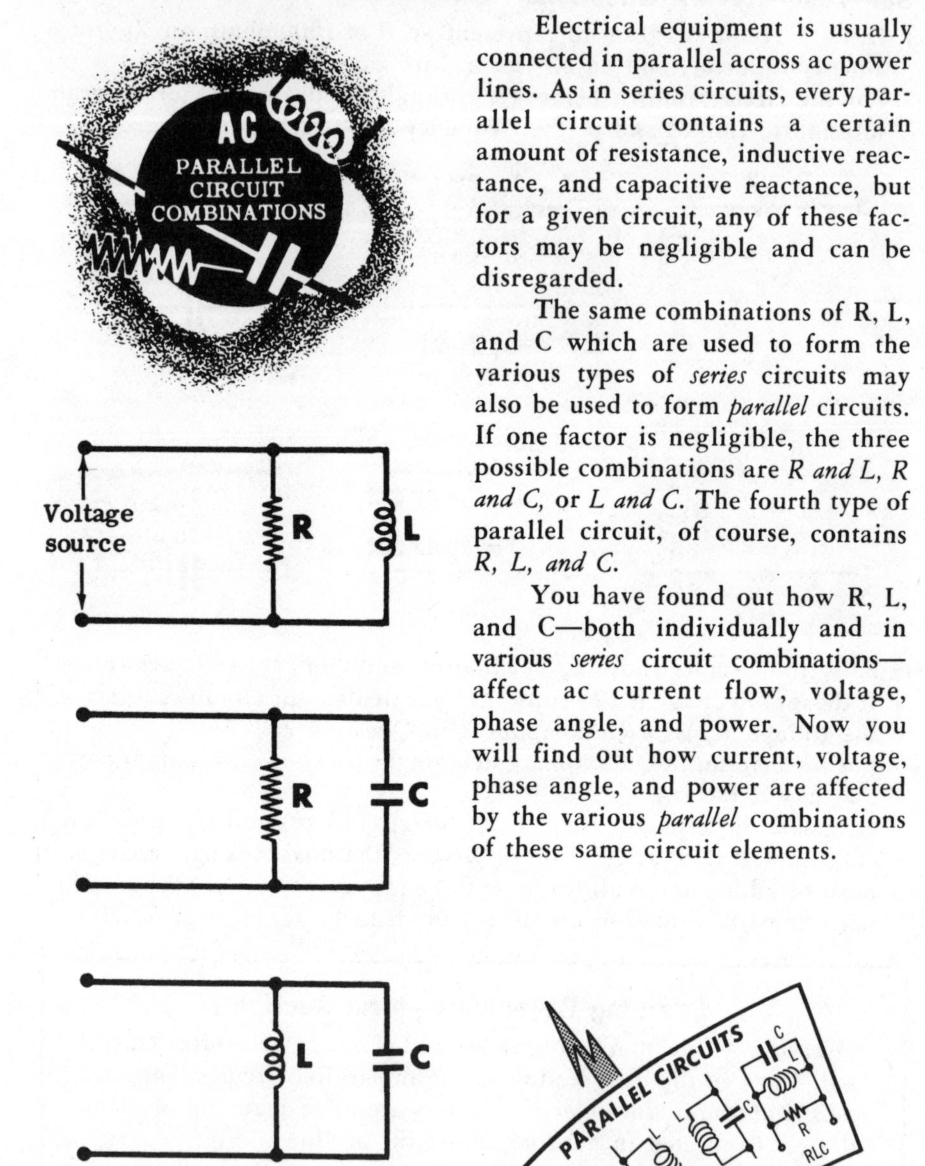

Electrical equipment is usually connected in parallel across ac power lines. As in series circuits, every parallel circuit contains a certain amount of resistance, inductive reactance, and capacitive reactance, but for a given circuit, any of these factors may be negligible and can be disregarded.

The same combinations of R, L, and C which are used to form the various types of *series* circuits may also be used to form *parallel* circuits. If one factor is negligible, the three possible combinations are *R and L, R and C,* or *L and C.* The fourth type of parallel circuit, of course, contains *R, L, and C.*

You have found out how R, L, and C—both individually and in various *series* circuit combinations—affect ac current flow, voltage, phase angle, and power. Now you will find out how current, voltage, phase angle, and power are affected by the various *parallel* combinations of these same circuit elements.

R L C

Voltages in AC Parallel Circuits

You will remember that in a parallel dc circuit the voltage across each of the parallel branches is *equal.* This is also true of ac parallel circuits; the voltages across each parallel branch are equal and also equal E_t , the total voltage of the parallel circuit. Not only are the voltages equal, but they are also *in phase.*

For example, if the various types of electrical equipment shown below—a lamp (resistance), a filter choke (inductance), and a capacitor (capacitance)—are connected in parallel, the voltage across each is *exactly the same.*

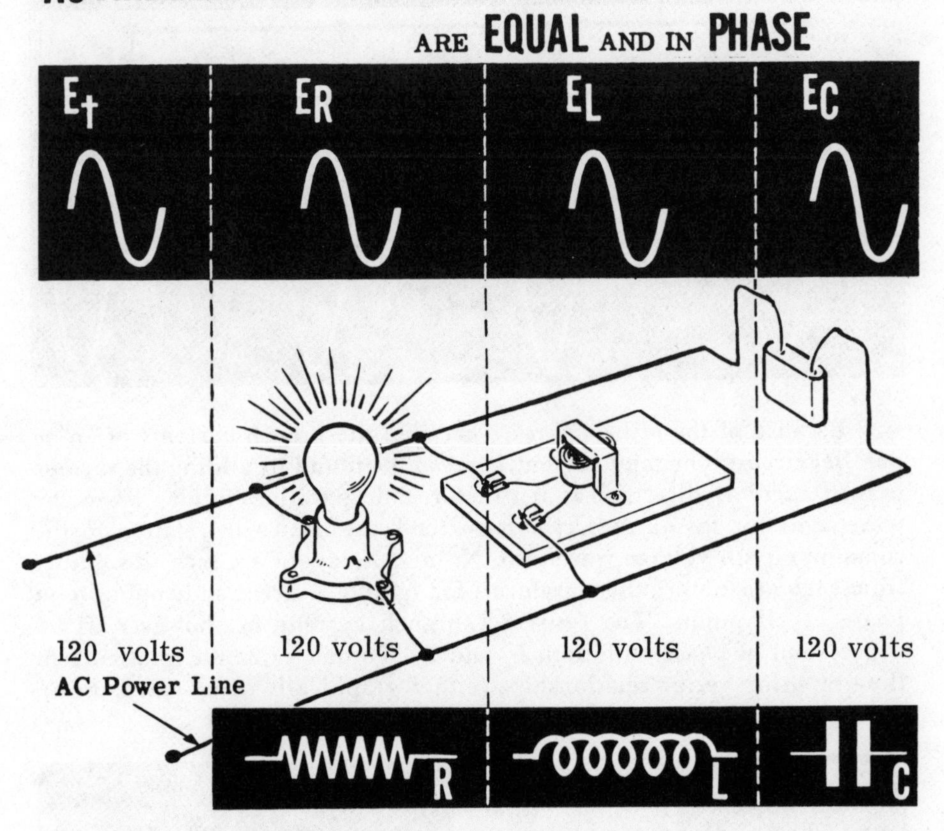

Regardless of the number of parallel branches, the value of the voltage across them is *equal* and *in phase.* All of the connections to one side of a parallel combination are considered to be one electrical point as long as the resistance of the connecting wire is neglected.

As you can see, in a parallel circuit the *voltages* across each element are the *same* while the *currents vary.* In the series circuits you studied earlier, on the other hand, the *current* through all elements is the *same,* but the *voltages* across the elements *vary.*

Currents in AC Parallel Circuits

The current flow through each individual branch is determined by the opposition offered by that branch. If your circuit consists of three branches—one a resistor, another an inductor, and the third a capacitor—the current through each branch depends on the resistance or reactance of that branch. The resistor branch current I_R is in phase with the circuit voltage E_t, while the inductor branch current I_L *lags* the circuit voltage by 90 degrees, and the capacitor branch current I_C *leads* the voltage by 90 degrees.

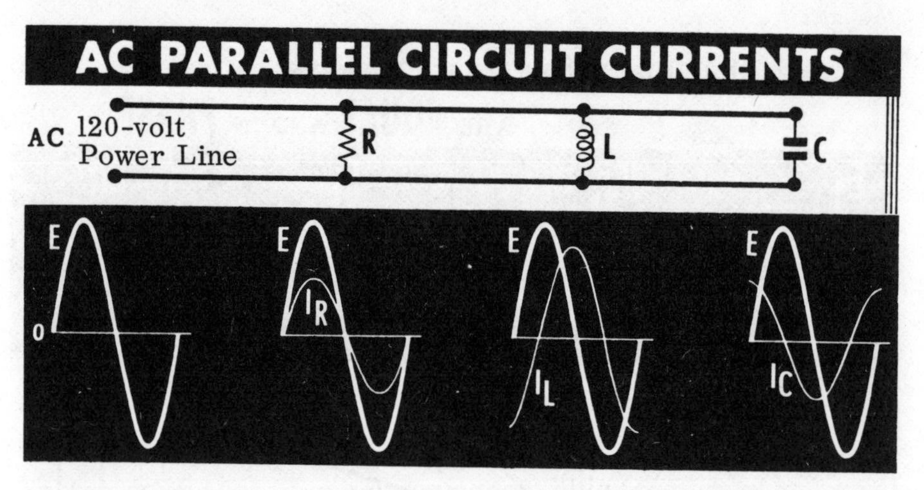

Because of the *phase difference* between the branch currents of an ac parallel circuit, the total current I_t *cannot* be found by adding the various branch currents directly—as it can for a dc parallel circuit. When the waveforms for the various circuit currents are drawn in relation to the common circuit voltage waveform, X_L and X_C again are seen to subtract from each other since the waveforms for I_L and I_C are exactly opposite in phase at all points. The resistance branch current I_R, however, is 90 degrees out of phase with both I_L and I_C. To determine the total current flow by using vector relationships (either graphically or by calculation), I_R must be combined with the difference between I_L and I_C.

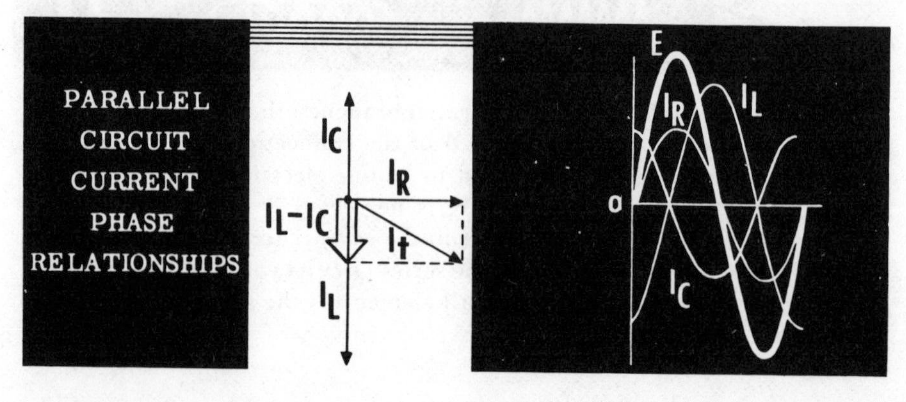

Currents in AC Parallel Circuits (continued)

To add the branch currents in an ac parallel circuit, the instantaneous values of current can be combined, as voltages were for series circuits, to obtain the instantaneous values of the total current waveform. After all the possible instantaneous values of current are obtained, the total current waveform is drawn by adding the instantaneous values at each point.

COMBINING PARALLEL CIRCUIT BRANCH CURRENTS

instantaneous total current is the sum of the three instantaneous values of I_R, I_C and I_L

I_C

TOTAL CURRENT I_t

I_R

I_L

A + B - C = D.

R L C

The maximum value of I_t is equal to or less than the sum of the maximum values of the individual currents and is not usually in phase with the various branch currents. With respect to the circuit voltage, the total current either *leads* or *lags* by a phase angle between 0 and 90 degrees, depending on whether the inductive or capacitive reactance is greater.

A vector diagram for the various circuit currents and the circuit voltage of an ac parallel circuit is similar to the vector diagram for circuit current and voltages for an ac series circuit. In the series circuit, voltages are drawn with reference to total circuit *current,* while for parallel circuits the different currents are drawn with reference to the total circuit *voltage.*

R and L Parallel Circuit Currents

If an ac parallel circuit consists of a resistance and inductance connected in parallel, and the circuit capacitance is negligible, the total circuit current is a combination of I_R (the current through the resistance) and I_L (the current through the inductance). I_R is in phase with the circuit voltage E_t, while I_L lags the voltage by 90 degrees.

PHASE RELATIONSHIPS IN AN R AND L PARALLEL CIRCUIT

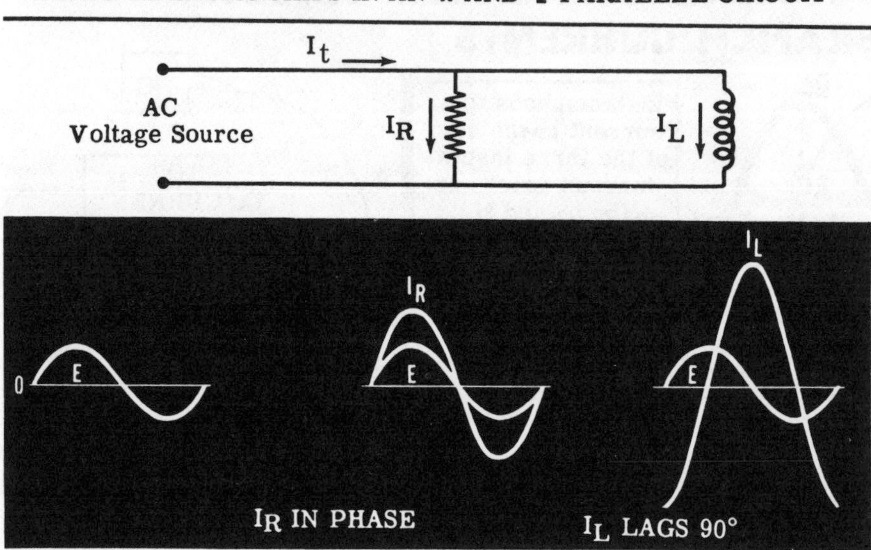

To find the total current I_t, you could draw I_R and I_L to scale and in the proper phase relationship to each other and then combine the corresponding instantaneous values to plot the total current waveform. This waveform then shows both the maximum value and phase angle of I_t.

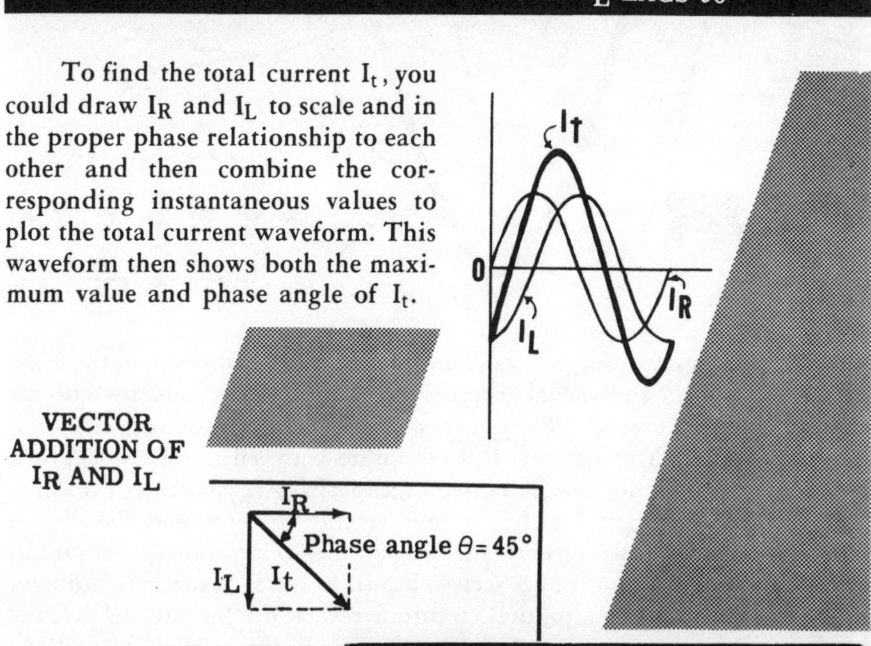

COMBINING WAVEFORMS I_R AND I_L

R and L Parallel Circuit Currents (continued)

You can use the much easier method of drawing vectors to scale to represent I_R and I_L, as you learned earlier, then combining the vectors by completing the parallelogram and drawing the diagonal, thus obtaining both the value and phase angle of I_t. The length of the diagonal represents the value of I_t, whereas the angle between I_t and I_R is the phase angle between total circuit voltage, E_t, and the total circuit current, I_t. As you know from right triangle relationships, the total current can be calculated from the Pythagorean theorem as

$$I_t = \sqrt{I_R^2 + I_L^2}$$

and the phase angle can be calculated as

$$\theta = \tan^{-1} \frac{I_L}{I_R}$$

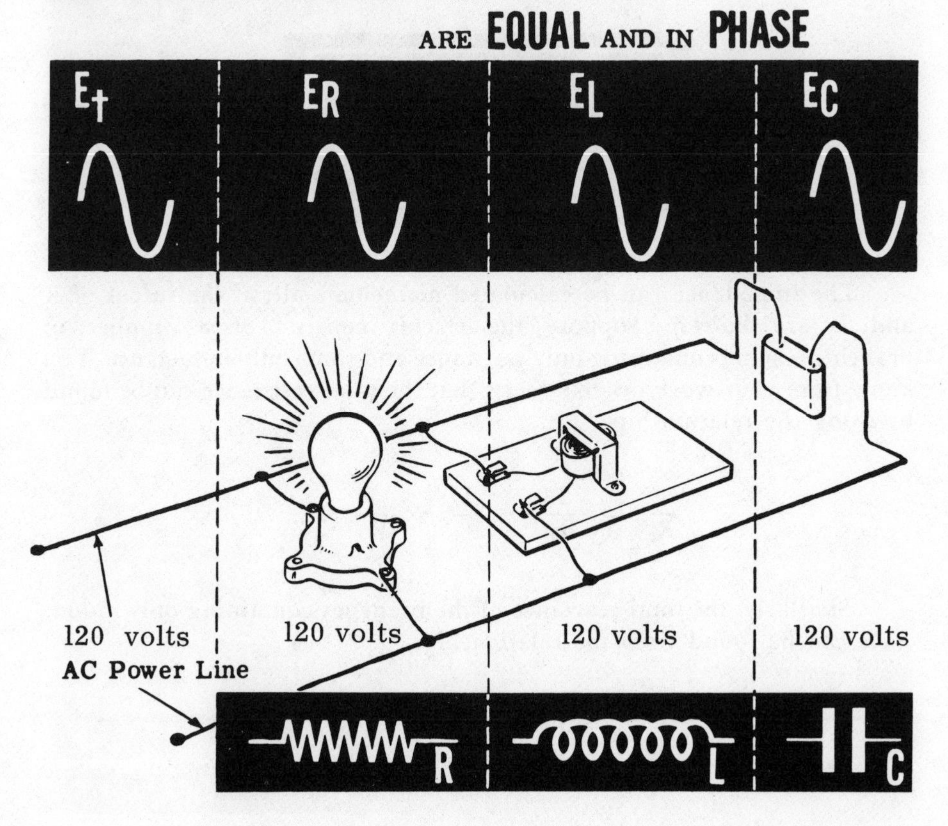

R and L Parallel Circuit Impedance

The impedance of a parallel circuit can be found by applying Ohm's law for ac current to the total circuit. Using Ohm's law, the impedance Z for all ac parallel circuits is found by dividing the circuit voltage by the total current, that is, $Z = E_t/I_t$.

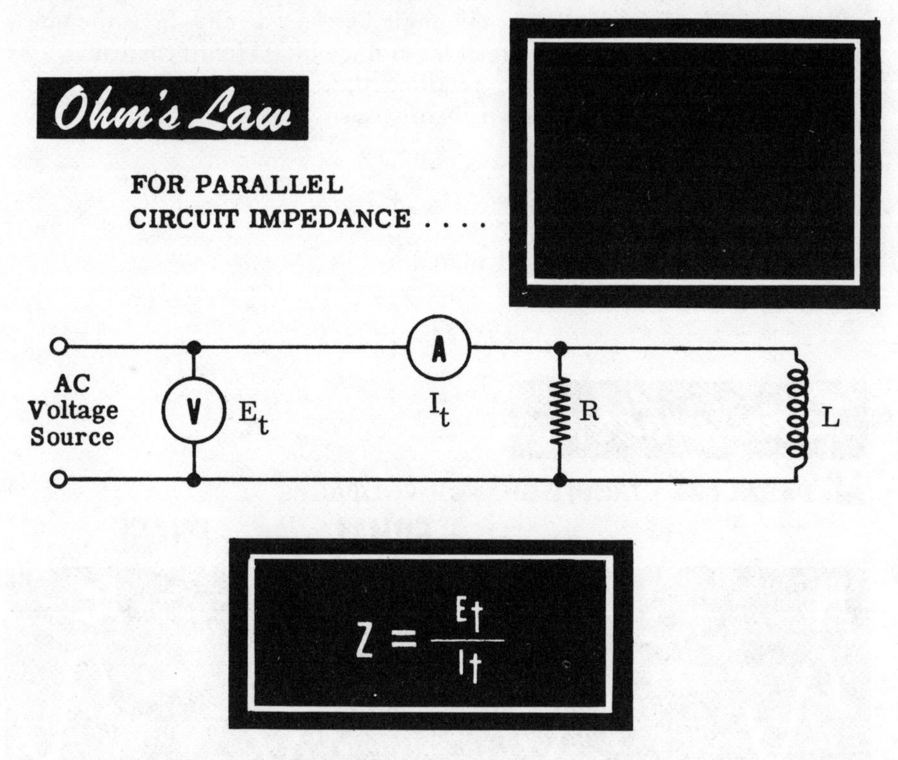

The impedance can be calculated mathematically if the values of R and L are known. Suppose the circuit consists of a number of branches—some containing only resistance and some only inductance. You know from your work on dc circuits that the total resistance can be found by using the relationship

$$\frac{1}{R_t} = \frac{1}{R_1} + \frac{1}{R_2} + \frac{1}{R_3} + \dots$$

Similarly, the total reactance of the branches containing only inductance can be found from the relationship,

$$\frac{1}{X_t} = \frac{1}{X_1} + \frac{1}{X_2} + \frac{1}{X_3} + \dots$$

R and L Parallel Circuit Impedance (continued)

The two quantities, R_t and X_t, can now be combined vectorially, as shown in the illustration below.

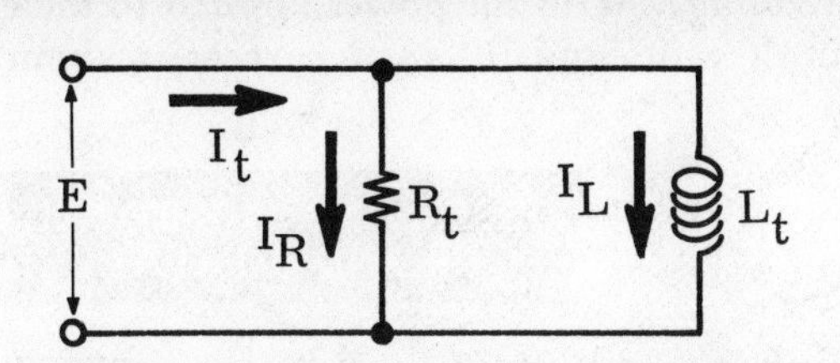

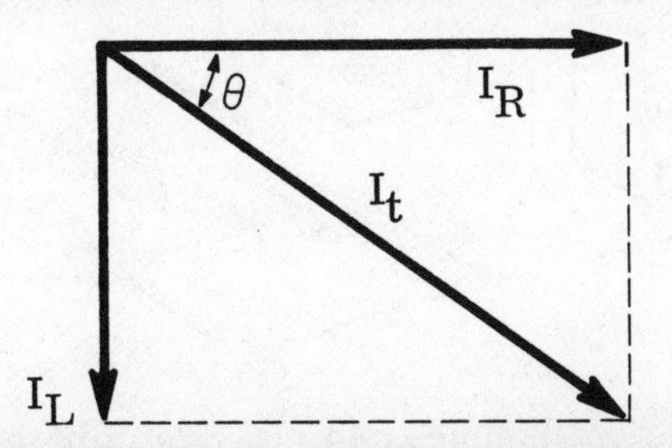

You will see that $I_R = E/R$, $I_L = E/X_L$, and $I_t = E/Z$, and since

$$I_t = \sqrt{I_R^2 + I_L^2}$$

we can say that

$$E/Z = \sqrt{(E/R)^2 + (E/X_L)^2}.$$

This can be rewritten

$$E/Z = E\sqrt{(1/R)^2 + (1/X_L)^2}.$$

Since E is on both sides of the equation, it can be deleted altogether, leaving

$$1/Z = \sqrt{(1/R)^2 + (1/X_L)^2}$$

so that

$$Z = 1/\sqrt{(1/R)^2 + (1/X_L)^2}$$

which can be simplified and expressed as

$$Z = RX_L/\sqrt{R^2 + X_L^2}$$

The phase angle can be calculated as always as $\theta = \tan^{-1}(I_L/I_R)$.

Experiment/Application—R-L Parallel Circuit Current and Impedance

To see for yourself the effects of connecting an inductor and a resistor in parallel, connect a 2,500-ohm, 20-watt resistor and a 5-henry inductor in parallel across the secondary of a step-down transformer to form an ac parallel circuit of R and L. A 0–50 mA ac milliammeter is connected to measure the total circuit current, and a 0–250 volt ac voltmeter is used to measure circuit voltage. With the power applied to the circuit, the circuit voltage is about 60 volts and the total current is about 40 mA.

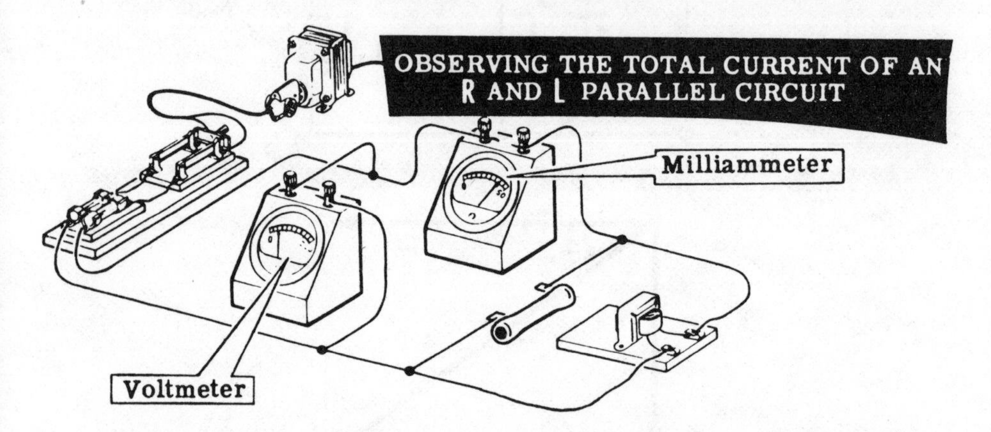

First connect the milliammeter to measure only the resistor current, then to measure only the inductor current. You see that the milliammeter reading for I_R is about 24 mA and that the current indicated for I_L is approximately 32 mA. The sum of these two branch currents I_R and I_L is thus 56 mA, while the actual measured total circuit current is about 40 mA. This difference shows that the branch currents must be added by means of vectors.

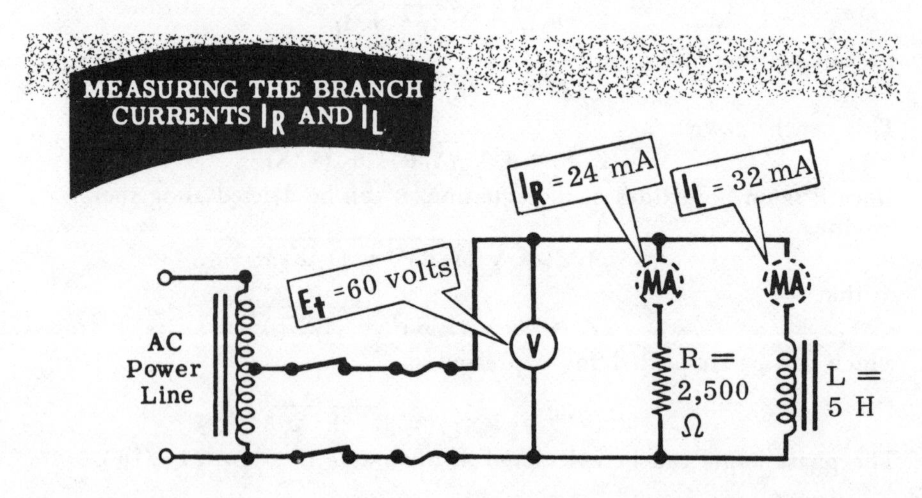

Experiment/Application—R-L Parallel Circuit Current and Impedance (continued)

Applying Ohm's law to the R and L circuit above to find the impedance, we get

$$Z = \frac{E}{I_t} = \frac{60}{0.040} = 1{,}500 \text{ ohms}$$

You can check the formula on page 4-57 by substituting the known values of R and X_L for this circuit and by comparing the value of Z found with that obtained above:

$$R = 2{,}500 \text{ ohms}; \quad X_L = 2\pi \times 60 \times 5 = 1{,}885 \text{ ohms}$$

$$Z = \frac{2{,}500 \times 1{,}885}{\sqrt{2{,}500^2 + 1{,}885^2}} = 1{,}500 \text{ ohms (approximately)}$$

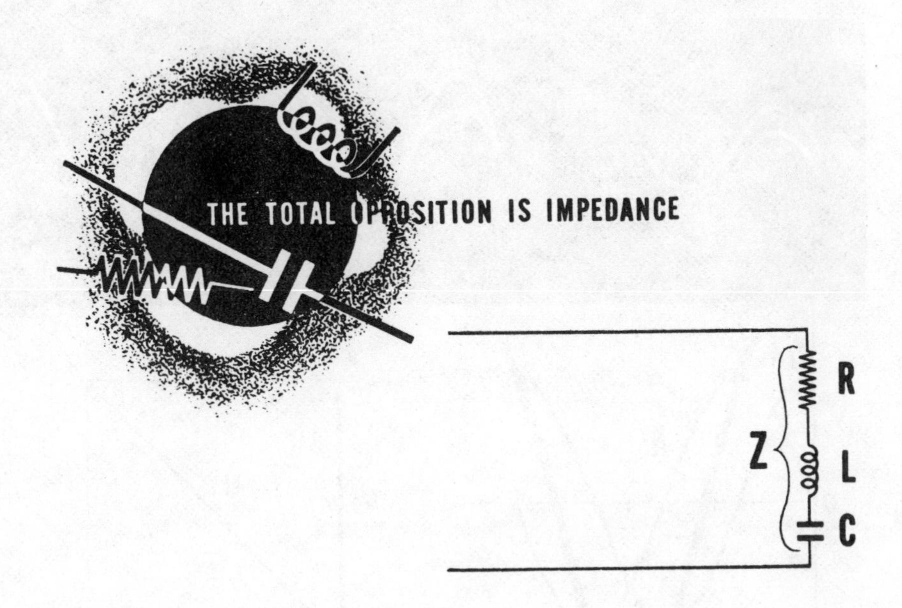

Ohm's Law for AC Circuits.

$$Z = \frac{E}{I_t} \text{ in AC circuits}$$

A
R
E
120 V
V
L
I

R and C Parallel Circuit Currents

The total current of an ac parallel circuit which consists only of R and C can be found by vectorially adding I_R (the resistive current) and I_C (the capacitive current). I_R is in phase with the circuit voltage E_t, while I_C leads the voltage by 90 degrees. To find the total current and its phase angle when I_R and I_C are known, you can draw either the waveforms of I_R and I_C or their vectors.

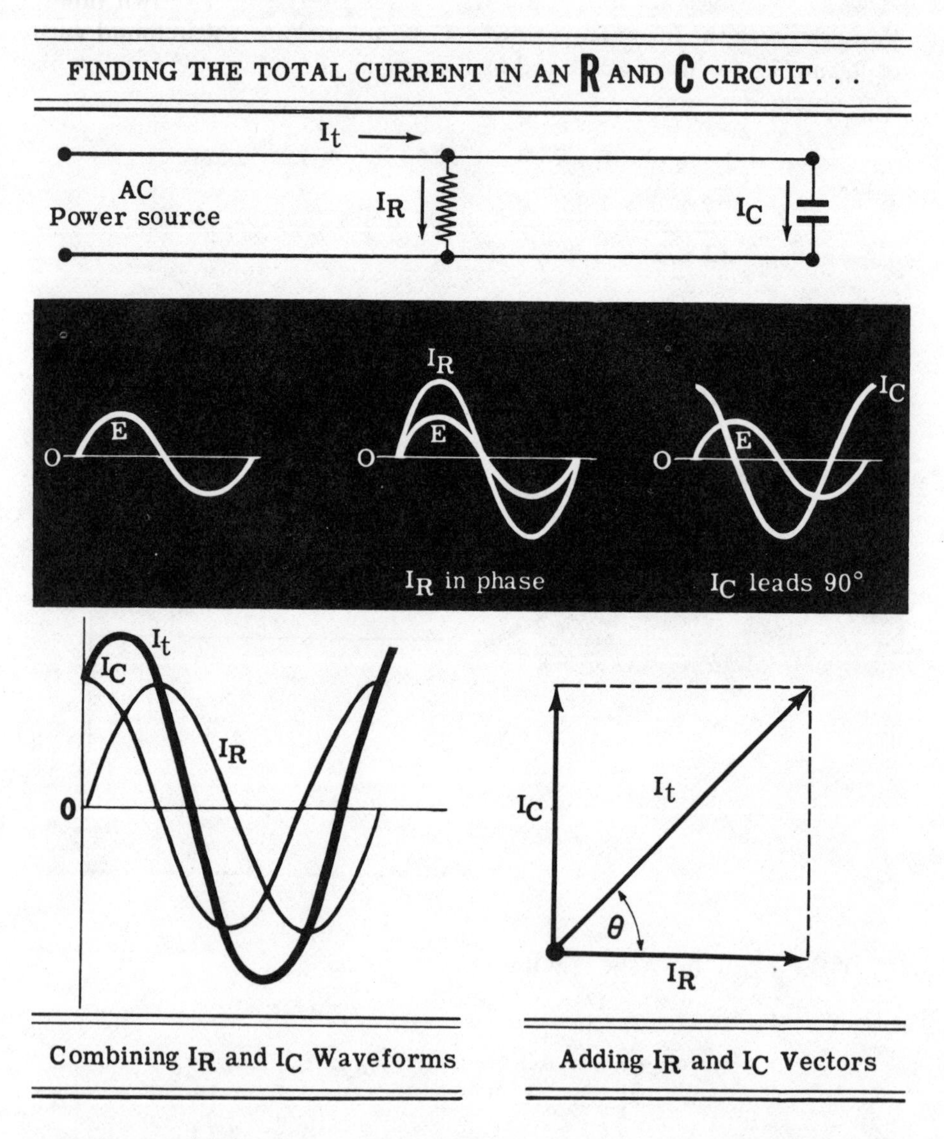

Combining I_R and I_C Waveforms

Adding I_R and I_C Vectors

As you did for R-L circuits, shown on page 4-54, you can find I_t by graphical addition or by calculation using the Pythagorean theorem with I_C in place of I_L.

R and C Parallel Circuit Impedance

The impedance in an R and C parallel circuit may be obtained by measuring the total current and voltage for the circuit and then finding Z by Ohm's law.

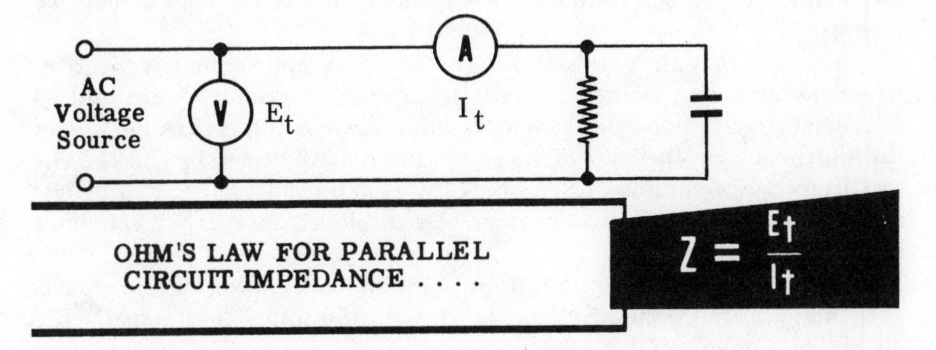

The impedance can also be calculated mathematically if the values R and C are known.

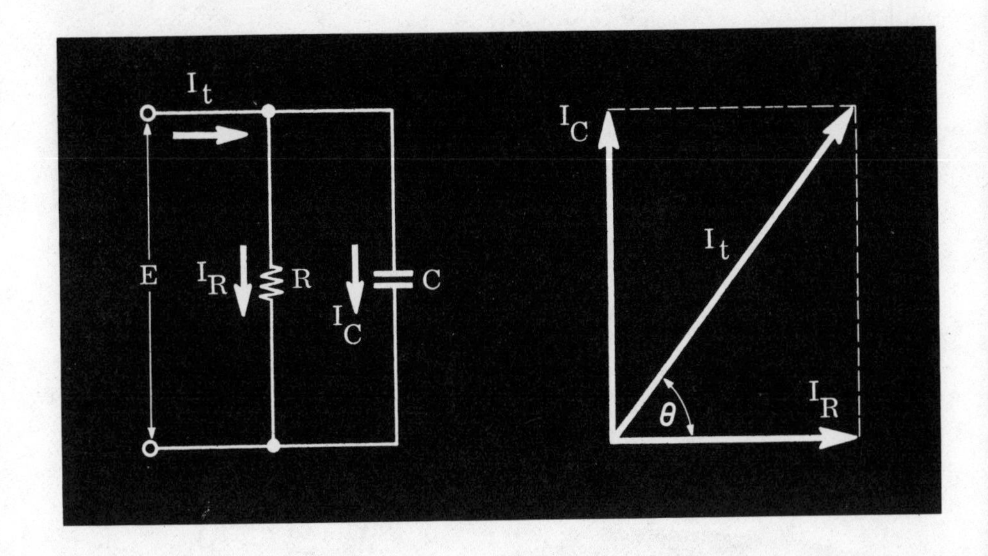

You will see that $I_R = E/R$, $I_C = E/X_C$ and $I_t = E/Z$ and since $I_t = \sqrt{I_R{}^2 + I_C{}^2}$, we can say that $E/Z = \sqrt{(E/R)^2 + (E/X_C)^2}$.

Now, if we eliminate E, as we did on page 4-57, we obtain $1/Z = \sqrt{(1/R)^2 + (1/X_C)^2}$ and $Z = RX_C/\sqrt{R^2 + X_C{}^2}$

As always, the phase angle can be calculated as $\theta = \tan^{-1}(X_C/I_R)$.

Experiment/Application—R-C Parallel Circuit Current and Impedance

Repeat the experiment/application described on page 4-58 using a 1-μF capacitor instead of the 5-henry inductor. First the total circuit voltage and current (E_t and I_t) are measured, and then the branch currents I_R and I_C.

You see that the total circuit current I_t is approximately 33 mA, while the measured branch currents I_R and I_C are about 24 mA and 23 mA, respectively. You also see that the total current is less than the sum of the branch currents because of the phase difference between I_R and I_C. The total impedance is about 1,820 ohms ($60 \div 0.033 = 1,818$), a value less than the opposition offered by either branch alone, since R = 2,500 ohms and X_C = 2,654 ohms.

Again, you can check the formula for the calculation of impedance, and compare the calculated impedance with the impedance actually obtained, by applying Ohm's law:

$$R = 2,500; \qquad X_C = \frac{10^6}{2\pi \times 60 \times 1} = 2,654 \text{ ohms}$$

$$Z = \frac{2,500 \times 2,654}{\sqrt{2,500^2 + 2,654^2}} = 1,820 \text{ ohms}$$

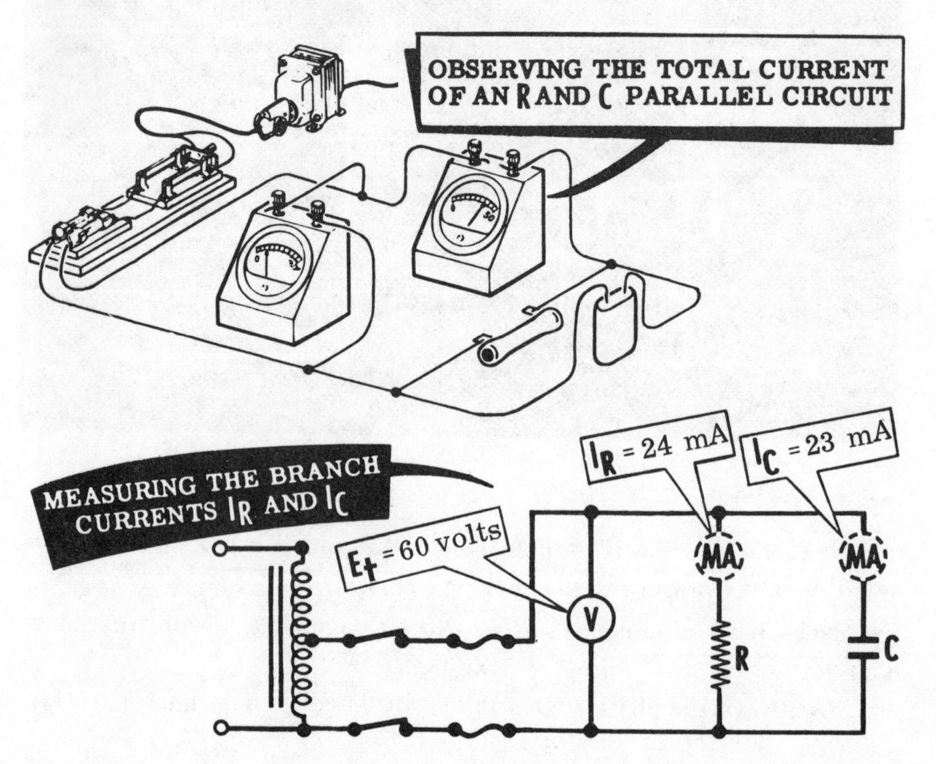

L and C Parallel Circuit Currents

When an ac parallel circuit consists only of L and C, the total current is equal to the *difference* between I_L and I_C, since they are *exactly opposite* in phase relationship. When the waveforms for I_L and I_C are drawn, you see that all the instantaneous values of I_L and I_C are of opposite polarity. If all the corresponding combined instantaneous values are plotted to form the waveform of I_t, the maximum value of this waveform is equal to the difference between I_L and I_C. For such circuits the total current can be found by subtracting the smaller current, I_L or I_C, from the larger.

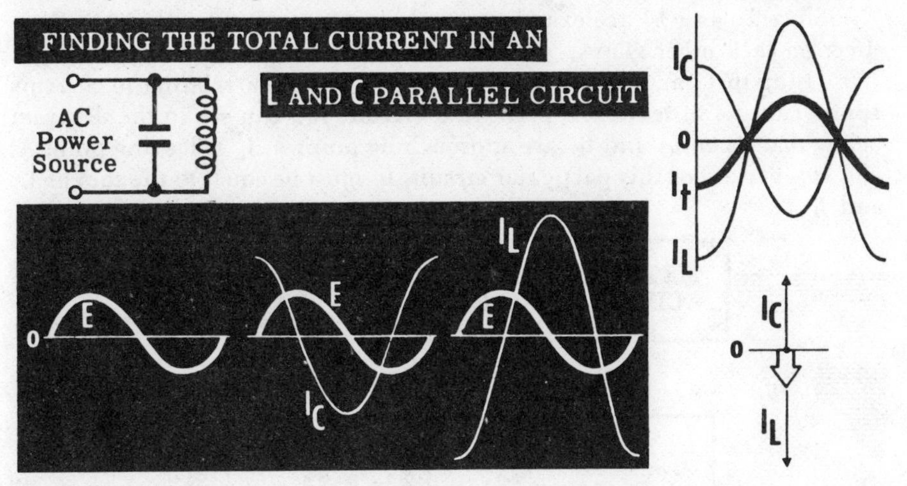

The relationships and paths of circuit currents in L and C circuits are shown below.

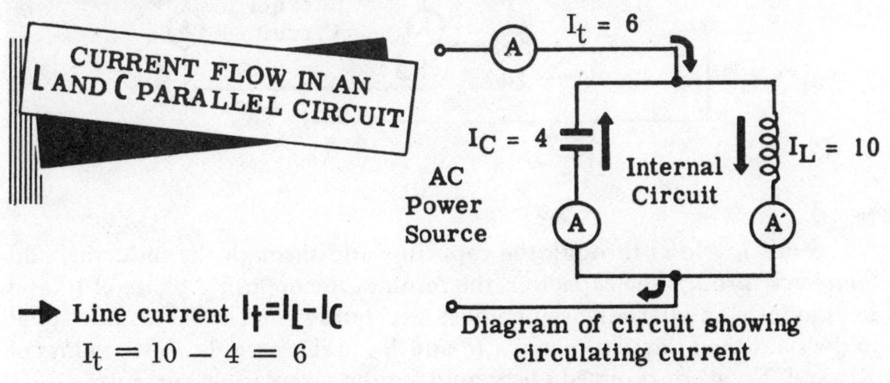

The parallel circuit can also be considered as consisting of an internal and an external circuit. Since the current flowing through the inductance is exactly opposite in polarity to that which is flowing through the capacitance at the same time, an internal circuit is formed. The amount of current flow around this internal circuit is equal to the smaller of the two currents I_L and I_C. The amount of current flowing through the external circuit (the voltage source) is equal to the difference between I_L and I_C.

L and C Parallel Circuit Currents and Impedance

The relationship between the various currents in a parallel circuit consisting of L and C is illustrated in the following example. A capacitor and an inductor are connected in parallel across a 60-Hz, 150-volt source so that $X_L = 50$ ohms and $X_C = 75$ ohms. The currents in the circuit are

$$I_L = \frac{E}{X_L} = \frac{150}{50} = 3\text{ A} \qquad I_C = \frac{E}{X_C} = \frac{150}{75} = 2\text{ A}$$

Since I_L and I_C are exactly opposite in phase, they have a cancelling effect on each other. Thus, the total current $I_t = I_L - I_C = 3 - 2 = 1$ A.

Using this phase relationship and Kirchhoff's law relating to currents approaching and leaving a point in a circuit, you can see in the diagram below that when I_C and I_t are approaching point A, I_L is leaving point A, and vice versa. For this particular circuit, I_L must be equal to the sum of I_t and I_C.

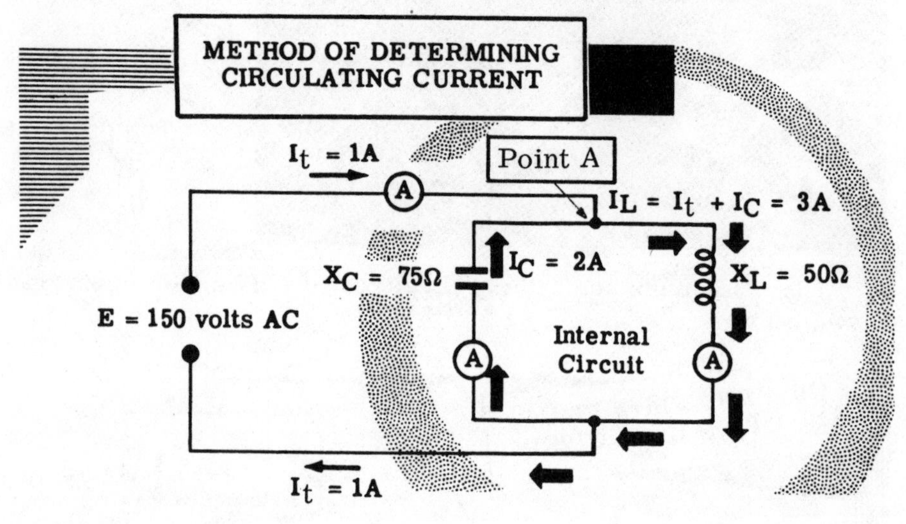

Since I_C flows through the capacitor and through the inductor, and then back through the capacitor, the result of the opposing phase of I_L and I_C is to form an internal circuit whose circulating current has a value equal to the smaller of the two currents I_L and I_C, in this case, I_C. If the values of X_L and X_C were reversed, I_L would be the circulating current.

The impedance of the circuit may be found by measuring the total current through the circuit and applied voltage, then by applying Ohm's law, or by calculation, if L and C are known. Since $I_t = I_L - I_C$,

$$\frac{E}{Z} = \frac{E}{X_L} - \frac{E}{X_C}$$

Hence,

$$\frac{1}{Z} = \frac{1}{X_L} - \frac{1}{X_C}$$

Experiment/Application—L-C Parallel Circuit Currents and Impedance

To observe the opposite effects of L and C in a parallel circuit, the 2,500-ohm resistor in the circuit of the experiment/application on page 4-62 is replaced by the 5-henry filter choke forming an L and C parallel circuit. Repeat each step of the experiment, first measuring the total circuit current, then that of each branch.

You see that the total circuit current is about 9 mA, while I_L is about 32 mA and I_C is about 23 mA. Thus, the total current is not only less than that of either branch but is actually the difference between I_L and I_C.

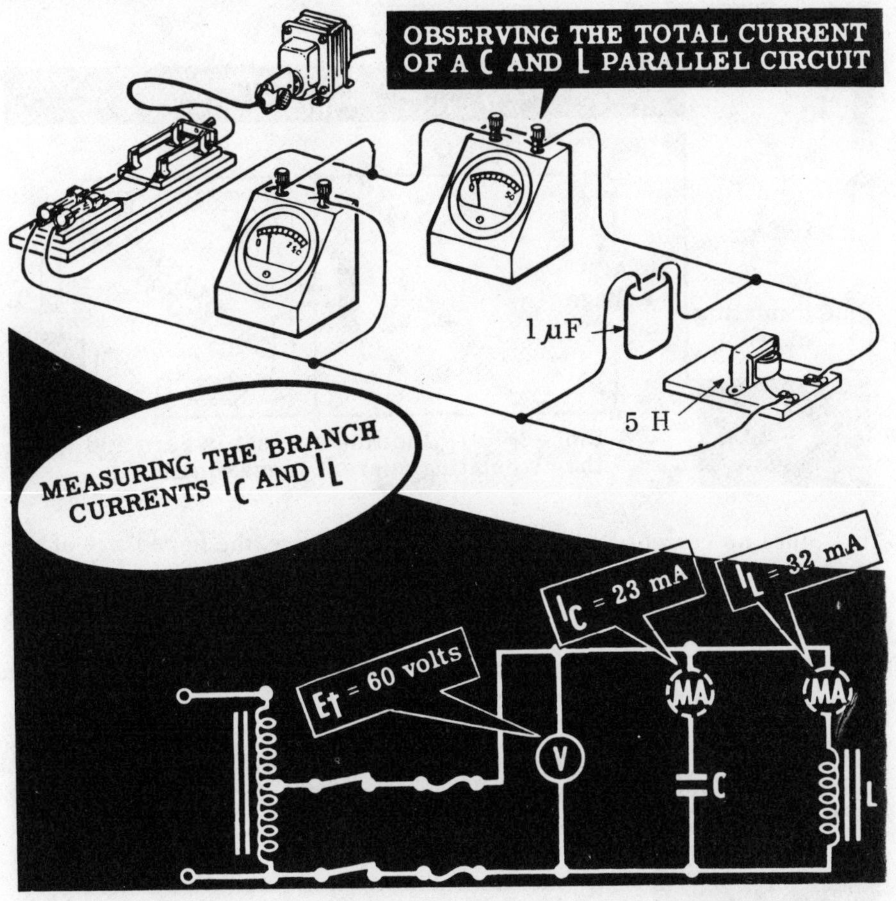

The total circuit impedance of the L and C circuit is 6,667 ohms ($60 \div 0.009 \approx 6{,}667$), a value greater than the opposition of either the L or C branch of the circuit ($X_L = 1{,}885$ ohms, $X_C = 2{,}650$ ohms). Notice that when L and C are both present in a parallel circuit, the impedance increases, an effect *opposite* to that of a series circuit where combining L and C results in a lower impedance. As you will learn in the next pages, at parallel resonance, the line current is zero for a pure L-C parallel circuit.

Parallel Circuit Resonance

In an L and C parallel circuit containing equal X_L and X_C, the external circuit current is equal to that flowing through any parallel resistance. If the circuit contains no resistance, the external current is zero. However, within a theoretical circuit consisting only of L and C with $X_L = X_C$, a large current called the *circulating current* will flow, using no current from the power supply. This happens because the corresponding instantaneous values of the currents I_L and I_C are always in opposite directions, and since the line current is the difference between I_L and I_C, if these values are equal, no external circuit current will flow. Such a circuit is called a *parallel resonant circuit.*

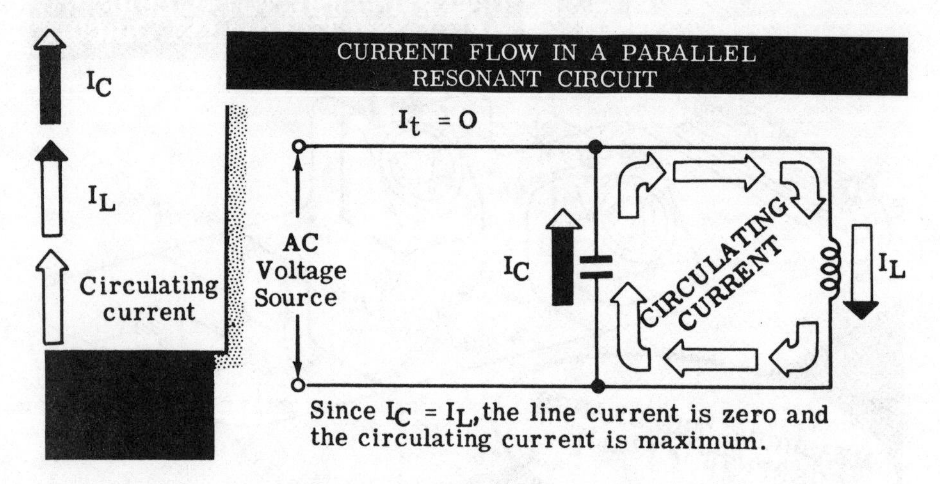

Since no current flows from the voltage source, the impedance of the ideal parallel resonant circuit is infinite. You will remember that in the case of the series circuit, the impedance at resonance was a minimum (and would be zero in a theoretical circuit).

Ohm's law for ac when applied to a parallel resonant circuit can be used to determine the value of the internal circulating current.

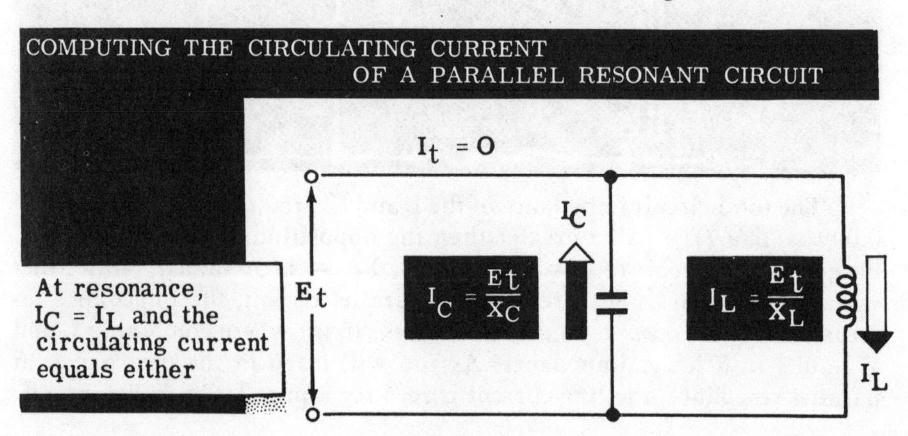

Parallel Circuit Resonance (continued)

Since the condition for parallel resonance is X_L being equal to X_C, the same as for series resonance, the same formula may be used to calculate parallel resonant conditions: $f_r = 1/2\pi\sqrt{LC}$.

As in the case of a series resonant circuit, if either the frequency, inductive reactance, or capacitive reactance of a circuit is varied and the two other values kept constant, the circuit current variation forms a resonance curve. However, the parallel resonance curve is the opposite of a series resonance curve. The series resonance current *increases* to a maximum at resonance, then *decreases* as resonance is passed, while the parallel resonance current *decreases* to a minimum at resonance, then *increases* as resonance is passed.

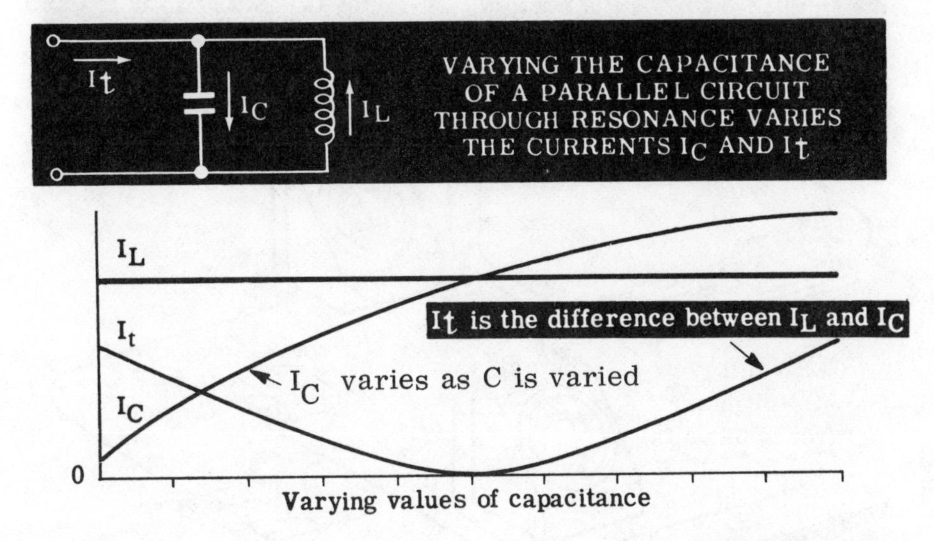

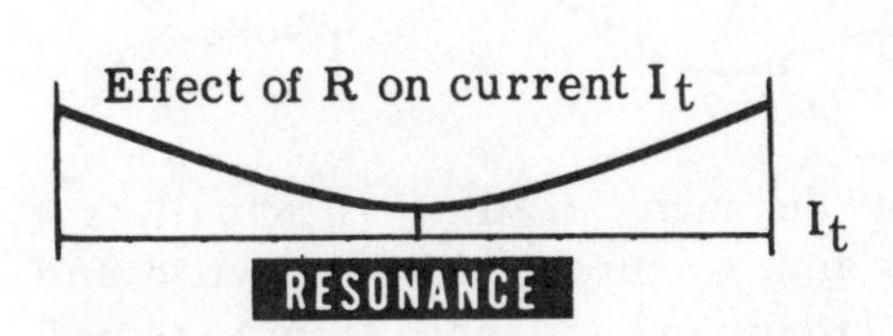

For a circuit of pure L and C, the curve would be as shown above. However, all actual capacitors and inductors have some resistance which prevents the current from becoming zero.

A comparison of circuit factors at resonance for series and parallel circuits, made in chart form, is shown below.

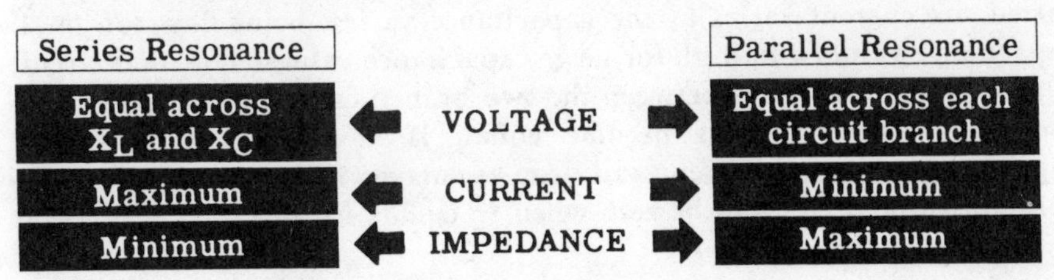

Series Resonance		Parallel Resonance
Equal across X_L and X_C	VOLTAGE	Equal across each circuit branch
Maximum	CURRENT	Minimum
Minimum	IMPEDANCE	Maximum

Experiment/Application—Parallel Circuit Resonance

To see the effect of parallel resonance, you could connect a 0.5-μF capacitor and a 5-henry inductor in parallel to form an L and C parallel circuit. A 0–50 mA ac milliammeter and a 0–250 volt ac voltmeter are connected to measure circuit current and voltage. This circuit is connected to the ac line through a switch, fuses, and step-down transformer. When the switch is closed, you observe that the current through the circuit is about 21 mA and that the voltage is approximately 60 volts.

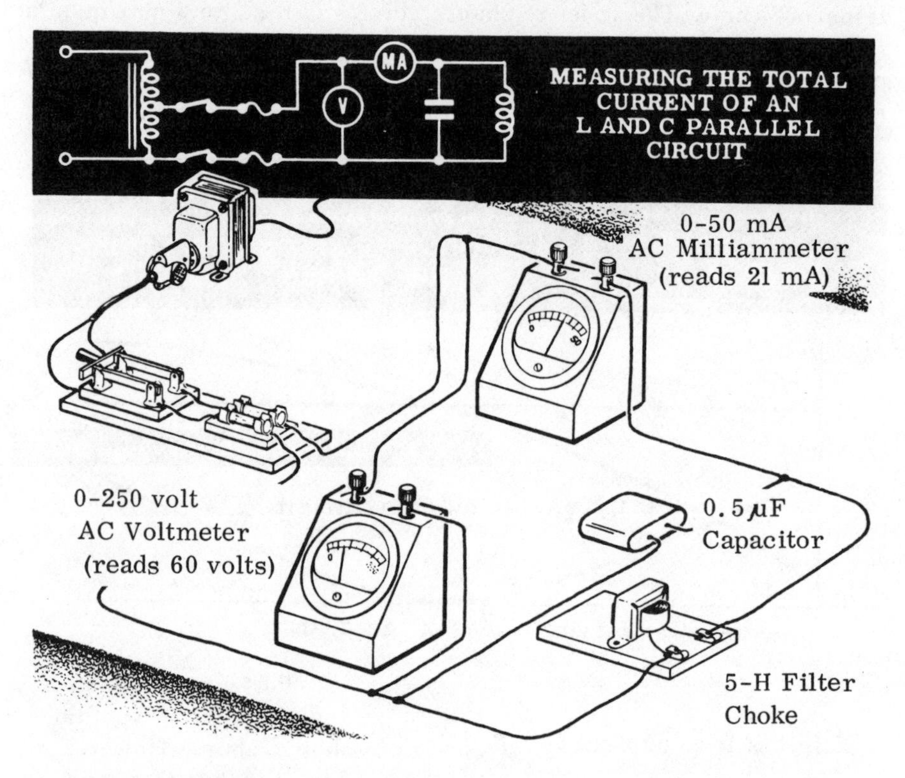

The total current indicated by the meter reading is actually the difference between the currents I_L and I_C through the inductive and capacitive branches of the parallel circuit (32 mA and 11 mA, respectively). The circuit voltage remains constant in parallel circuits so that, if a fixed value of inductance is used as one branch of the circuit, the current in that branch remains constant. If the capacitance of the other branch is varied, its current varies as the capacitance varies, being low for small capacitance values and high for large capacitance values. The total circuit current is the difference between the two branch currents; it is zero when the two branch currents become equal. If you increase the circuit capacitance, the total current will drop as current I_C increases toward the constant value of I_L, will be zero when I_C equals I_L, and then will rise as I_C becomes greater than I_L.

Experiment/Application—Parallel Circuit Resonance (continued)

Now vary the circuit capacitance in steps of 0.5 μF from 0.5 μF to 3.5 μF. Observe that the current decreases from approximately 21 mA to a minimum value less than 10 mA, then rises to a value beyond the range of the milliammeter. The current at resonance does not reach zero because the circuit branches are not purely capacitive and inductive and cannot be so in a practical parallel circuit. You will observe that the voltage does not change across either the branches or the total parallel circuit as the capacitance value is changed.

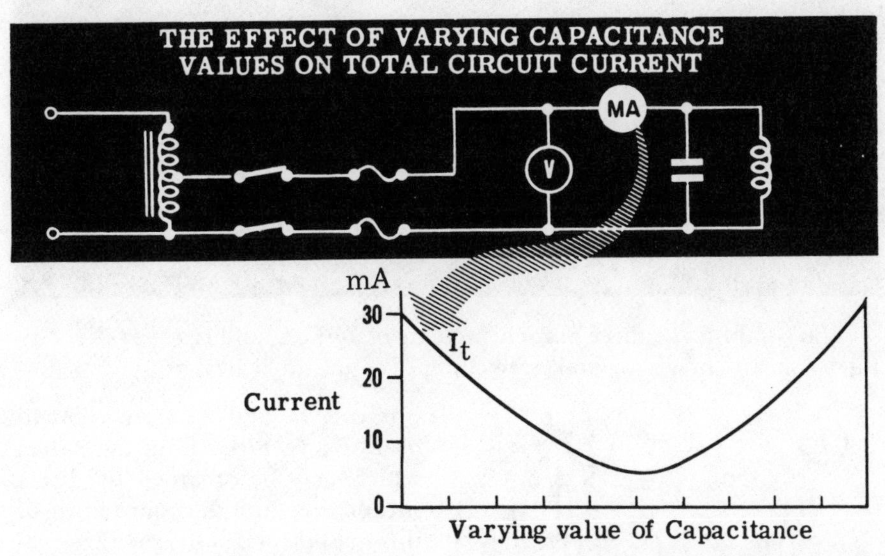

After the value of capacitance has been varied through the complete range of values, the value which indicates resonance—minimum current flow—is used to show that the circulating current exceeds the total current at resonance. Measure again the total current of the parallel resonant circuit; then connect the milliammeter to measure only the current in the inductive branch. You will see that the total current is less than 10 mA, yet the circulating current ($I_L = I_C$) is approximately 32 mA.

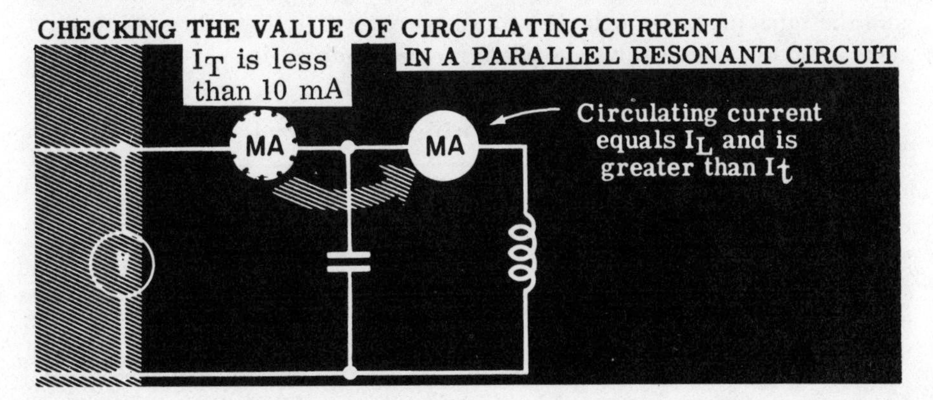

R, L, and C Parallel Circuit Currents

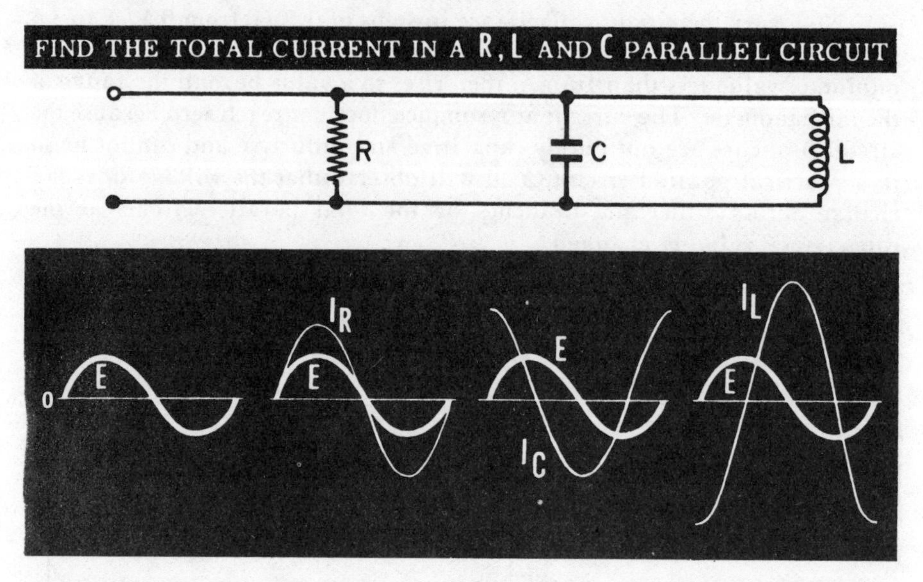

To combine the three branch currents of an R, L, and C ac parallel circuit by means of vectors takes two steps, as outlined below:

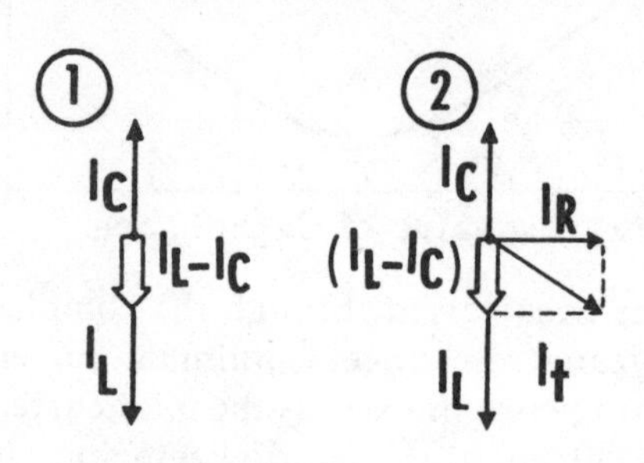

1. Currents I_L and I_C are combined by using vectors. (Both the value, which may be obtained by direct subtraction, and the phase angle of this combined current are required).
2. The combined value of I_L and I_C is then combined with I_R to obtain the total current.

In an R, L, and C parallel circuit—as in the parallel L and C circuit—a circulating current equal to the smaller of the two currents I_L and I_C flows through an internal circuit consisting of the inductance branch and the capacitance branch. The total current which flows through the external circuit (the voltage source) is the combination of I_R and the difference between currents I_L and I_C.

$$I_t = \text{Vector addition of } I_R + (I_L - I_C)$$

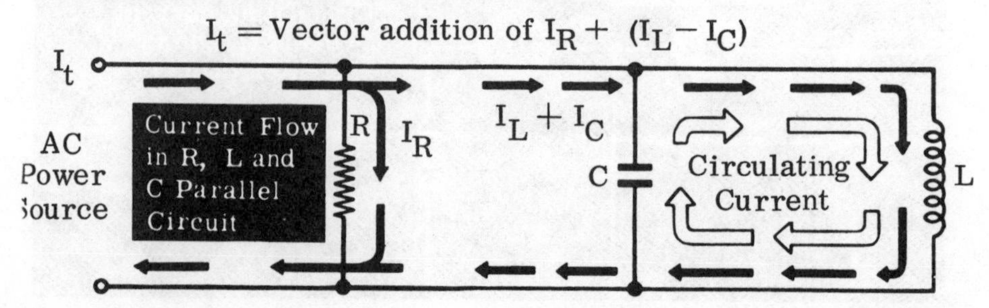

R, L, and C Parallel Circuit Impedance

The total impedance of an R, L, and C parallel ac circuit may be found by first measuring the current through the circuit and the applied voltage and then using Ohm's law.

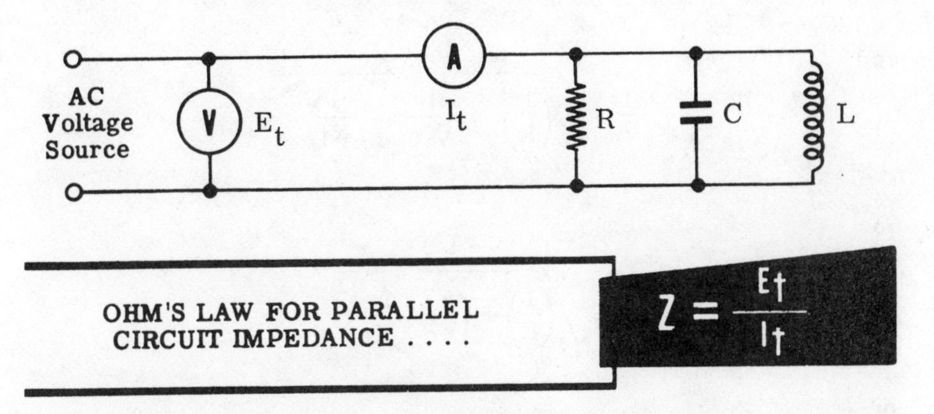

The impedance may be calculated mathematically (with the aid of a hand calculator) if the values of R, L, and C are known.

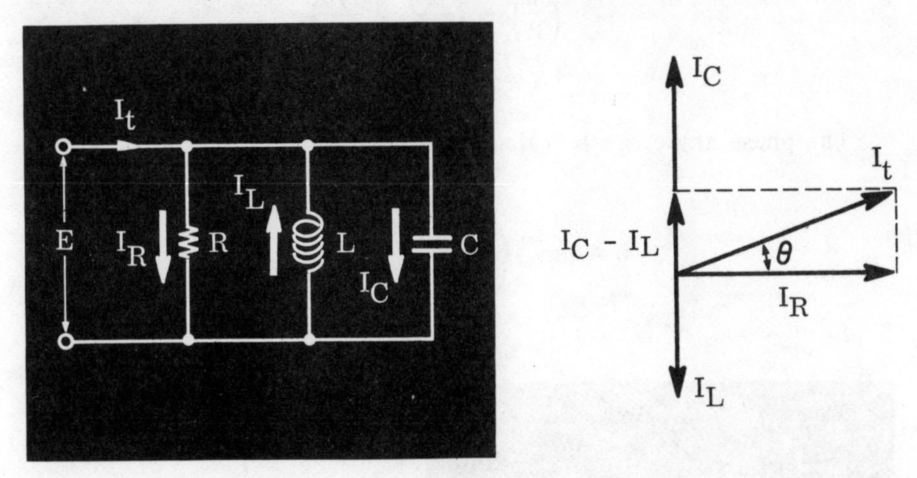

$$I_R = \frac{E}{R},\ I_L = \frac{E}{X_L},\ I_C = \frac{E}{X_C},\ \text{and } I_t = \frac{E}{Z}$$

$I_t = \sqrt{I_R^2 + (I_C - I_L)^2}$ in the example illustrated above, where the capacitive reactance (X_C) is smaller than the inductive reactance (X_L). But $I_t = \sqrt{I_R^2 + (I_L - I_C)^2}$ if the inductive reactance is smaller than the capacitive reactance.

The total current will lead the applied voltage if X_C is smaller than X_L and will lag the applied voltage if X_L is smaller than X_C, using the procedures of page 4-57.

R, L, and C Parallel Circuit Impedance (continued)

$$\frac{E}{Z} = \sqrt{\left(\frac{E}{R}\right)^2 + \left(\frac{E}{X_C} - \frac{E}{X_L}\right)^2}$$

and

$$\frac{1}{Z} = \sqrt{\left(\frac{1}{R}\right)^2 + \left(\frac{1}{X_C} - \frac{1}{X_L}\right)^2}$$

So

$$Z = \frac{1}{\sqrt{\left(\frac{1}{R}\right)^2 + \left(\frac{1}{X_C} - \frac{1}{X_L}\right)^2}}$$

or

$$\frac{1}{\sqrt{\left(\frac{1}{R}\right)^2 + \left(\frac{1}{X_L} - \frac{1}{X_C}\right)^2}}$$

The phase angle can be calculated as before as

$$\theta = \tan^{-1}\left(\frac{I_C - I_L}{I_R}\right)$$

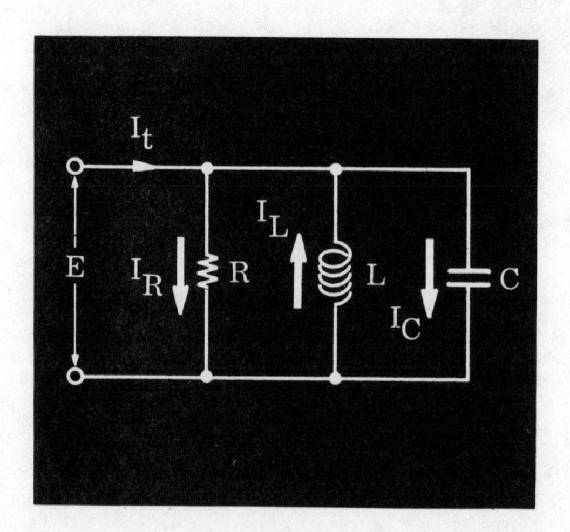

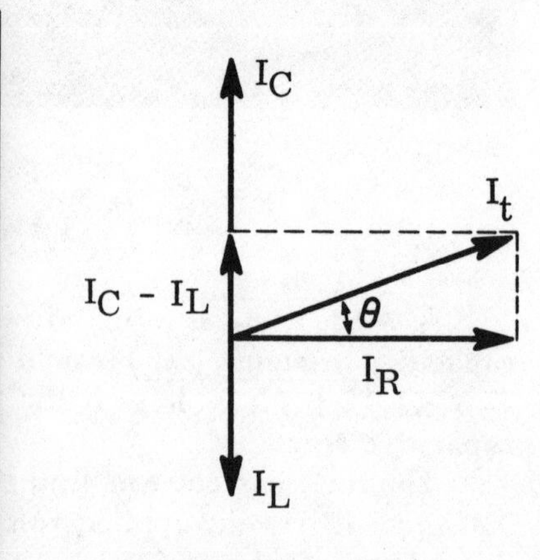

Experiment/Application—R, L, and C Parallel Circuit Currents and Impedance

By connecting a 2,500-ohm, 20-watt resistor in parallel with a 5-henry inductor and a 1-μF capacitor, you can make an R, L, and C parallel circuit. To check the various currents and find the total circuit impedance, measure the total circuit current and then the individual currents through the resistor, inductor, and capacitor in turn.

You will see that I_R is 24 mA, I_L is 31.8 mA, and I_C is 22.6 mA. Again you will see that the sum of the individual currents is much greater than the actual measured total current of 25.7 mA. The total circuit current is the sum of the resistive current I_R and the combined inductive and capacitive currents I_L and I_C added vectorially.

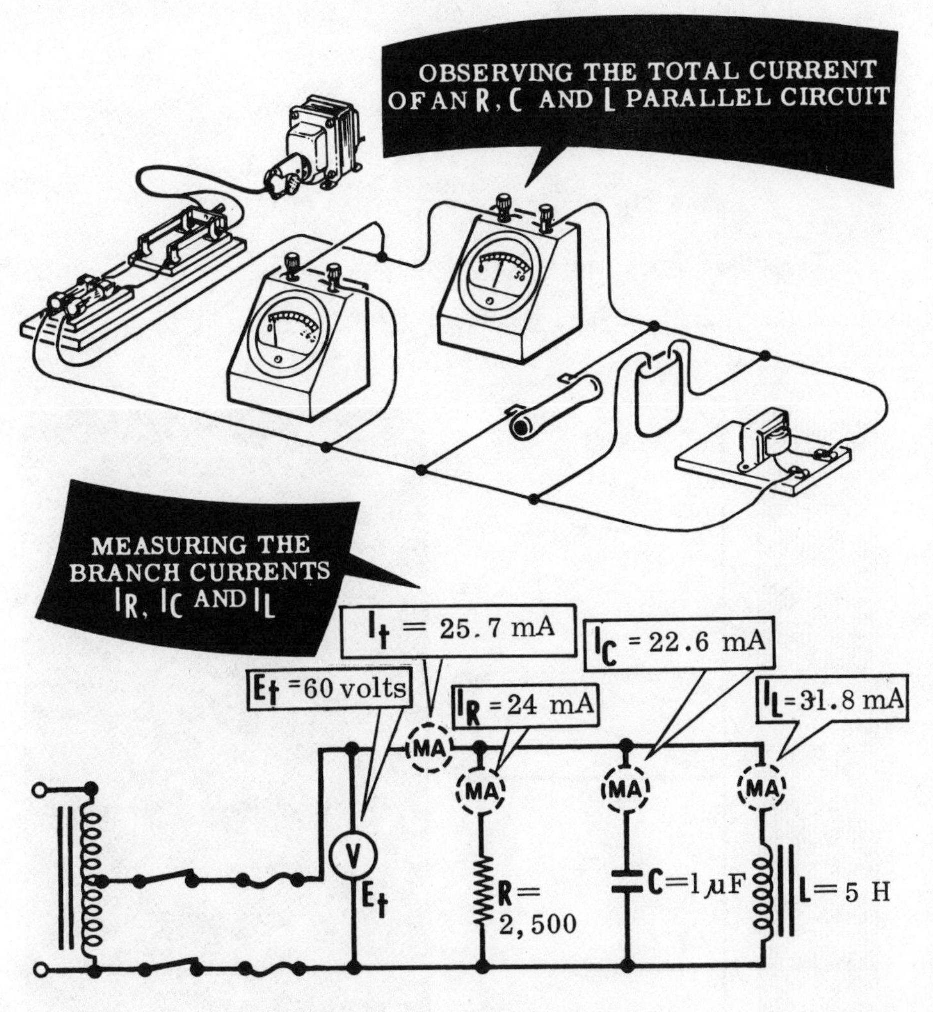

The total circuit impedance is about 2,335 ohms (60 ÷ 25.7 = 2,335).

Experiment/Application—R, L, and C Parallel Circuit Currents and Impedance (continued)

The impedance found by Ohm's law can be checked against the value found by substituting for R, X_L, and X_C in the formula on page 4-72:

$$Z = \frac{1}{\sqrt{\left(\frac{1}{R}\right)^2 + \left(\frac{1}{X_L} - \frac{1}{X_C}\right)^2}} = \frac{1}{\sqrt{\frac{1}{2{,}500^2} + \left(\frac{1}{1{,}885} - \frac{1}{2{,}650}\right)^2}}$$

$$= 2{,}335 \text{ ohms}$$

The circuit currents are:

$$I_L = \frac{E_t}{X_L} = \frac{60}{1{,}885} = 31.8 \text{ mA}$$

$$I_C = \frac{E_t}{X_C} = \frac{60}{2{,}650} = 22.6 \text{ mA}$$

$$I_R = \frac{E_t}{R} = \frac{60}{2{,}580} = 24 \text{ mA}$$

The phase angle can be calculated as:

$$\theta = \tan^{-1} \frac{I_L - I_C}{I_R} = \tan^{-1} \frac{31.8 - 22.6}{24} = 21.39^\circ$$

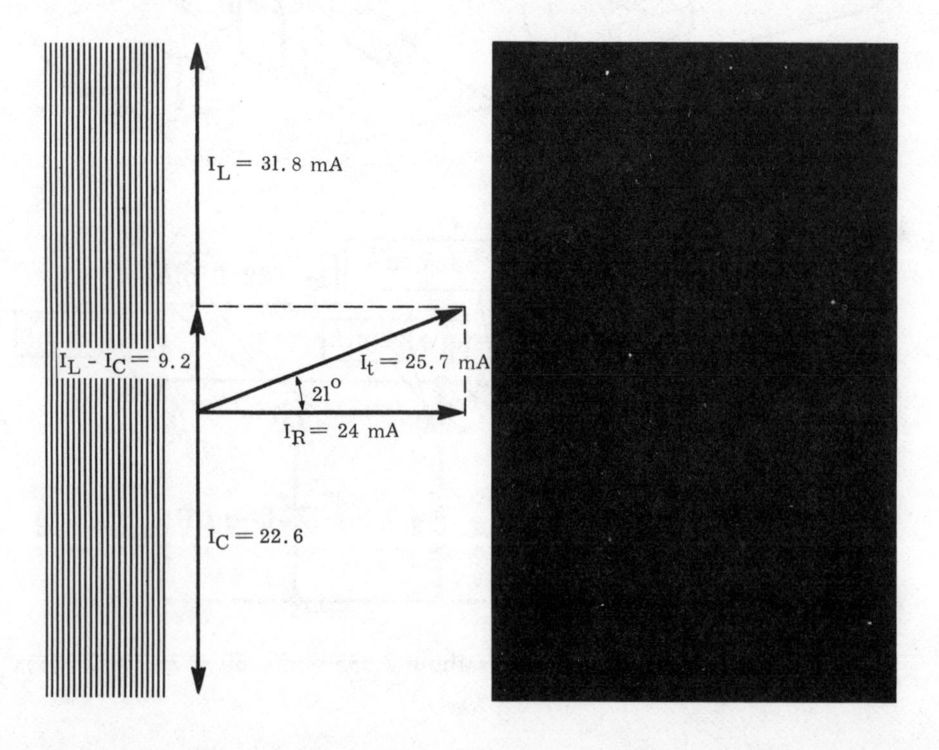

Power and Power Factor in Parallel Circuits

It would be wise at this point to review pages 4-44 to 4-46 on power and power factor in series circuits since the calculations are the *same* as for the parallel circuits we are now studying.

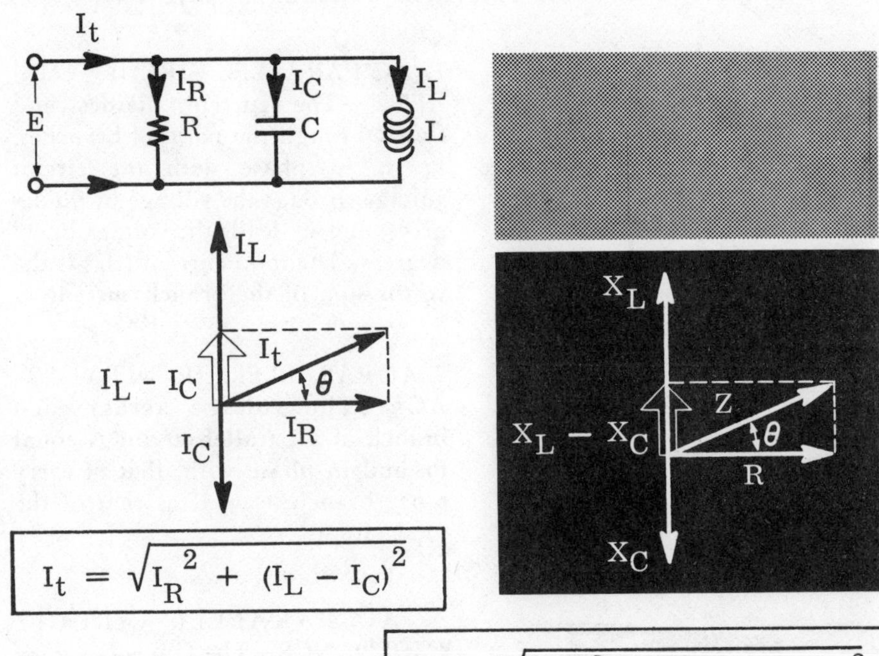

$$Z = 1/\sqrt{(1/R)^2 + (1/X_L - 1/X_C)^2}$$

$$\text{True power} = EI_R$$

$$\text{Apparent power} = EI_t$$

$$\text{Power factor} = \frac{\text{True power}}{\text{Apparent power}} = \frac{EI_R}{EI_t} = \frac{I_R}{I_t}$$

$$\text{Power factor} = \frac{R}{Z} = \frac{I_R}{I_t}$$

As before, the power factor is also defined as cos θ since

$$\cos\theta = \frac{R}{Z} \text{ or } \frac{I_R}{I_t}$$

$$\text{True power} = EI_t \cos\theta$$

As you can see, the relationships for power and power factor in parallel circuits are the *same* as those for series circuits.

Review of Current, Voltage, Impedance, Resonance, Power, Power Factor

Consider what you have found out so far about ac parallel circuits, and compare the effects of parallel connections of R, L, and C in ac circuits with those of the series connections reviewed on page 4-48.

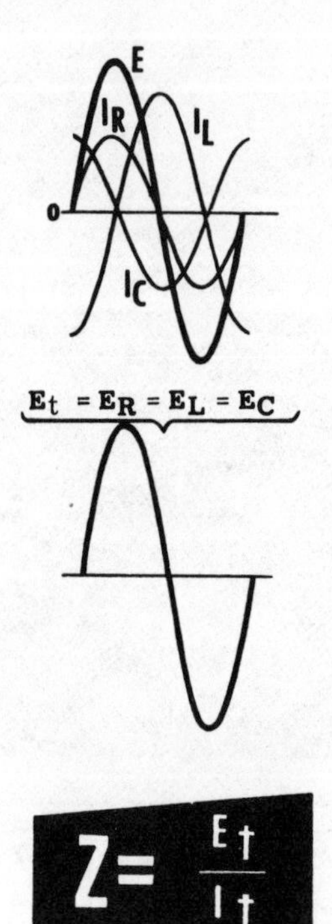

1. AC PARALLEL CIRCUIT CURRENT—The current divides and flows through the parallel branches. I_R is in phase with the circuit voltage, I_L lags the voltage by 90 degrees, and I_C leads the voltage by 90 degrees. The total current (I_t) is the vector sum of the branch currents.

2. AC PARALLEL CIRCUIT VOLTAGE—The voltage across each branch of a parallel circuit is equal to, and in phase with, that of every other branch as well as that of the total circuit.

3. AC PARALLEL CIRCUIT IMPEDANCE—The impedance of an ac parallel circuit is equal to the applied voltage divided by the total circuit current.

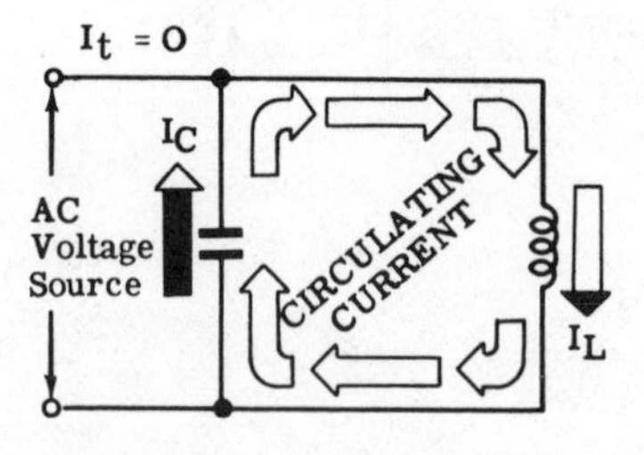

4. PARALLEL RESONANCE TOTAL CURRENT—The total current is minimum in a parallel resonant circuit. The circulating current is maximum in a parallel resonant circuit. At parallel resonance, the power factor is unity, and the phase angle is zero.

$$\text{True power} = I^2R = EI\cos\theta$$

$$\text{Apparent power} = EI$$

$$\text{Power factor} = \frac{\text{True power}}{\text{Apparent power}}$$

$$\text{Power factor} = \cos\theta = R/Z$$

5. POWER AND POWER FACTOR—As with all R, L, and C circuits, both true and apparent power are found in parallel circuits.

Self-Test—Review Questions

1. Draw a vector diagram to represent an R of 300 ohms, an X_L for a 1-henry inductor, and a 5-μF capacitor at 60 Hz.
2. Draw vector diagrams to solve the circuits shown below, finding both Z and θ (f = 60 Hz).

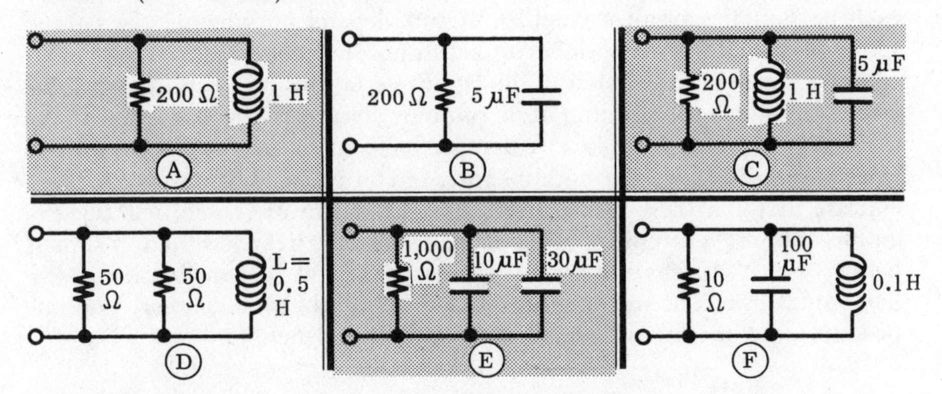

3. Solve the circuits above by calculation, and compare your answers.
4. If the input voltage is 120 volts, 60 Hz, calculate the circuit currents for each component and the total current in the six circuits of Question 2. Calculate the circuit currents where applicable.
5. For each circuit in Question 2, calculate true power, apparent power, and power factor.
6. Calculate the resonant frequency for circuits (e) and (f) of Question 2. How much current will flow at resonance? What is the phase? What capacitance must be added to circuit (c) to be resonant at 60 Hz? What inductance must be paralleled with the inductance of circuit (f) to make it resonant at 300 Hz?

Learning Objectives—Next Section

Overview—Now that you know about *series* and *parallel* ac circuits, you are ready to learn about *complex* ac circuits that have *both* series and parallel parts. You will find, as is true of dc complex circuits, that the learning trick is to reduce (simplify) the circuit by applying what you know about series and parallel circuit elements.

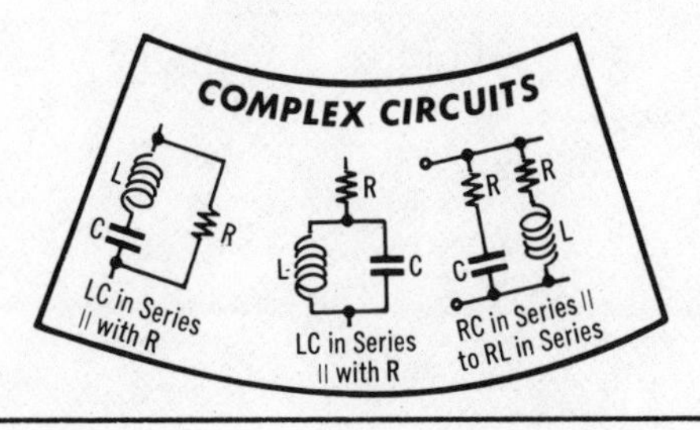

Series-Parallel Circuits

Many ac circuits are neither series circuits nor parallel circuits, but a *combination* of these two basic circuits. Such circuits are called *series-parallel* or *complex circuits.*

The values and phase relationships of the voltages and currents for each particular part of a complex circuit depend on whether the part is series or parallel. Any number of series-parallel combinations can form complex circuits. Regardless of the circuit variations, the step-by-step solution is similar to the solution of dc complex circuits that you learned about in Volume 2. The difference is that we must solve the circuits by vectors since we have to consider both magnitude and phase. The parts of the circuit are first considered separately; then the results are combined. For example, suppose a circuit consists of the series-parallel combination shown below, with two separate series circuits connected in parallel across the 240-volt ac line. The vector solution used to find the total current, total impedance, and the circuit phase angle is outlined below.

Series-Parallel Circuit

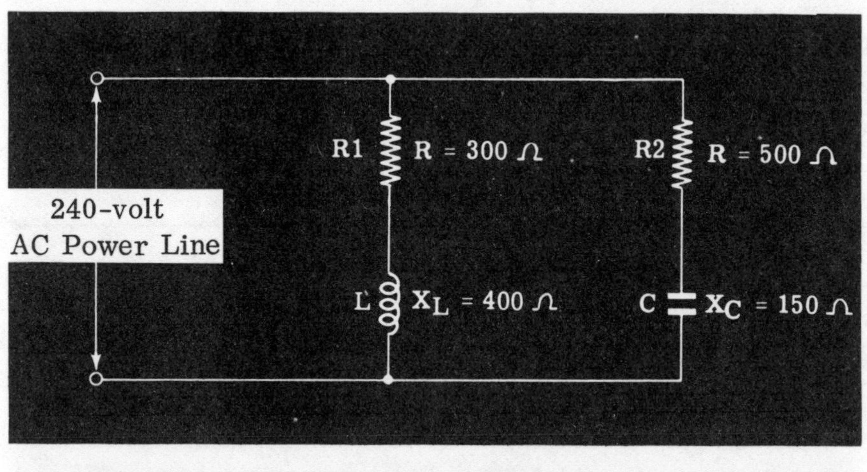

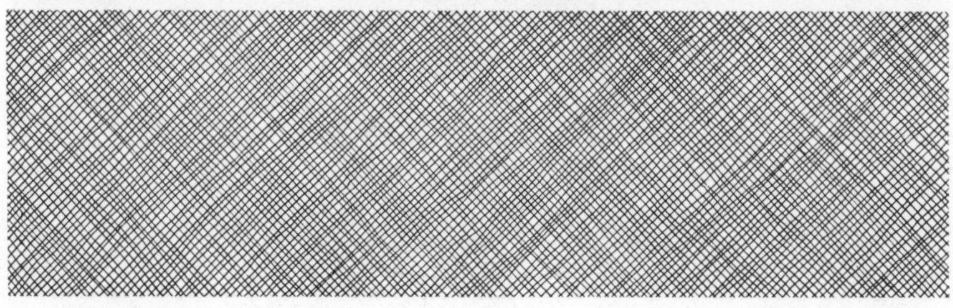

Series-Parallel Circuits (continued)

To find the values of branch currents I1 and I2, the impedance of each branch is found separately by using vectors. The current values are then determined by applying Ohm's law to the branches separately.

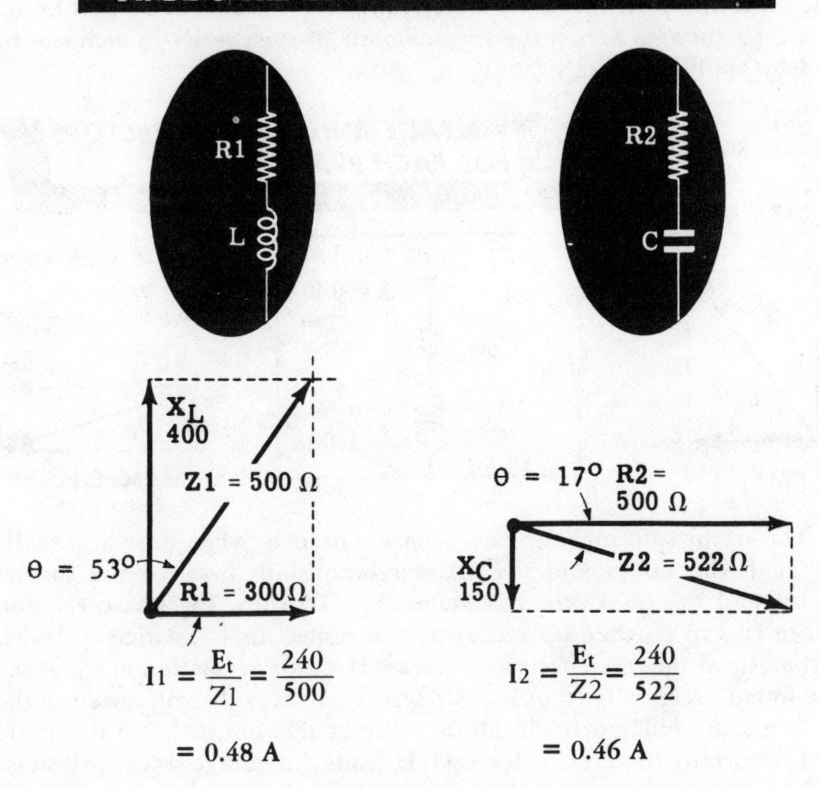

$$I1 = \frac{E_t}{Z1} = \frac{240}{500} = 0.48\ \text{A} \qquad I2 = \frac{E_t}{Z2} = \frac{240}{522} = 0.46\ \text{A}$$

You can, of course, calculate these circuits analytically.

$$Z1 = \sqrt{300^2 + 400^2} = 500\ \text{ohms} \qquad Z2 = \sqrt{500^2 + 150^2} = 522\ \text{ohms}$$

The phase angles are:

$$\theta = \tan^{-1}\frac{X_L}{R} = \tan^{-1}\frac{400}{300} = 53.1^\circ \qquad \theta = \tan^{-1}\frac{X_C}{R} = \tan^{-1}\frac{150}{500} = 16.7^\circ$$

Series-Parallel Circuits—Vector Graphic Solution

Although you know branch currents I1 and I2, the total current I_t *cannot* be found by adding I1 and I2 *directly*. Since they are *out of phase*, the total current must be found by *vector addition*.

To find the phase relationship between I1 and I2 so that they may be added by using vectors, the voltage and current vectors for each series branch must first be drawn separately. (Since the values of I1 and I2 are known, the voltages across the various parts of each series branch can be found by applying Ohm's law.)

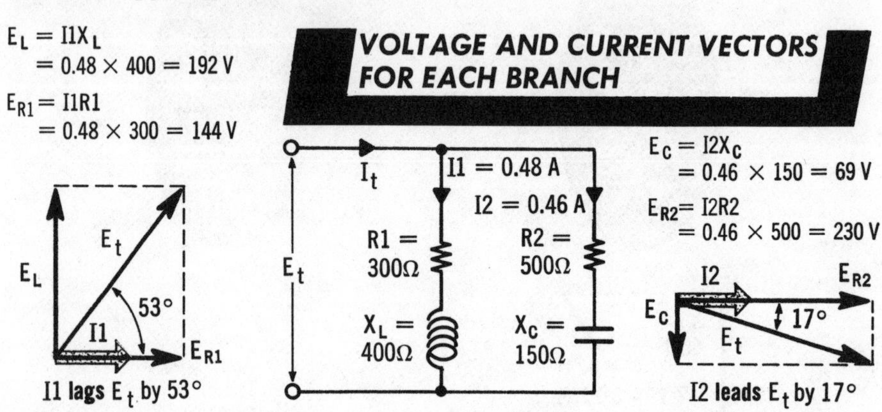

The vector solutions for each separate branch, when drawn to scale, show both the values and the phase relationships between the branch currents and the total circuit voltage, E_t. To show the phase relation between I1 and I2, they are redrawn with respect to E_t, which is drawn horizontally as the reference vector. Draw I1 down in relation to E_t at the angle found vectorially (that is, 53 degrees). I1 lags the voltage since the branch has the inductor in it. Draw I2 up in relation to E_t at the angle found vectorially (that is, 17 degrees). I2 leads the voltage since the branch has the capacitor in it. Complete the parallelogram by drawing a dotted line from the end of each vector parallel to the other. From the reference point, draw a line to where the two dotted lines cross. This vector represents the total current of the circuit. Measure with a protractor the angle between I_t and E_t, and this will be the phase angle of the circuit.

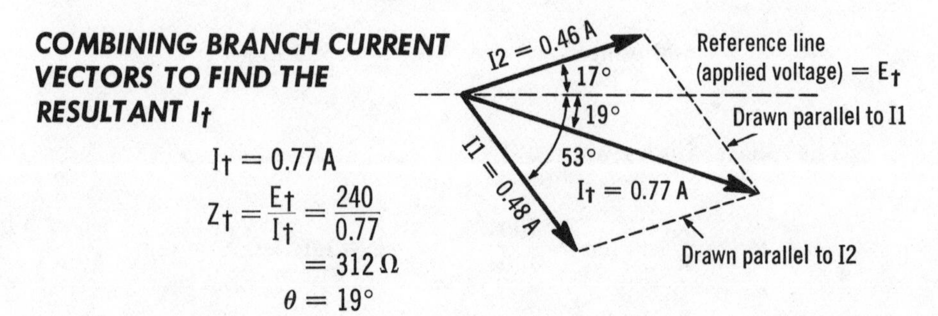

Series-Parallel Circuits—Solution by Calculation

Alternatively, the branch currents, total current, and total impedance of the circuit on page 4-80 can be calculated mathematically. Considering each branch separately,

$$Z1 = \sqrt{R1^2 + X_L^2} \quad \text{and} \quad Z2 = \sqrt{R2^2 + X_C^2}$$

$$Z1 = \sqrt{300^2 + 400^2} = 500 \text{ ohms} \quad \text{and}$$

$$Z2 = \sqrt{500^2 + 150^2} = 522 \text{ ohms}$$

The current through Z1, the inductive branch, lags behind the applied voltage by a phase angle $\theta 1$. $\theta 1 = \tan^{-1} X_L / R1 = 53$ degrees.

The current through the capacitance branch, Z2, leads the applied voltage by a phase angle $\theta 2$. $\theta 2 = \tan^{-1} X_C / R2 = 17$ degrees.

I1, the current through the inductive branch, is E/Z1, or 240/500 = 0.48 ampere.

I2, the current through the capacitive branch, is E/Z2, or 240/522 = 0.46 ampere.

As you know, both I1 and I2 can be resolved into two components—one component in phase with the applied voltage and the other 90 degrees out of phase with it.

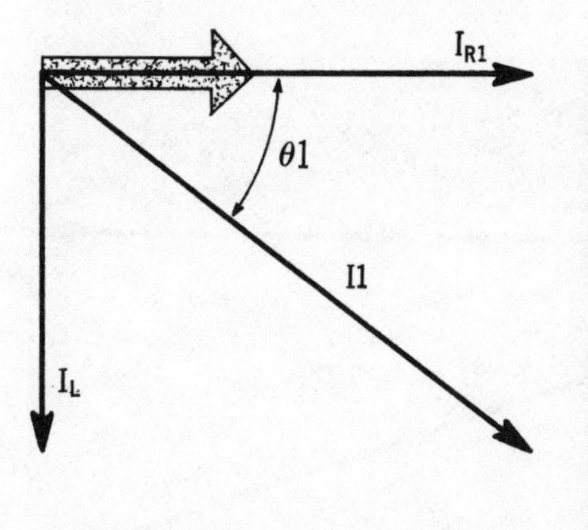

$I_{R1} = I1 \cos \theta 1 = 0.48 \cos 53° = 0.29$ A

$I_L = I1 \sin \theta 1$ (lagging) $= 0.48 \sin 53° = 0.38$ A

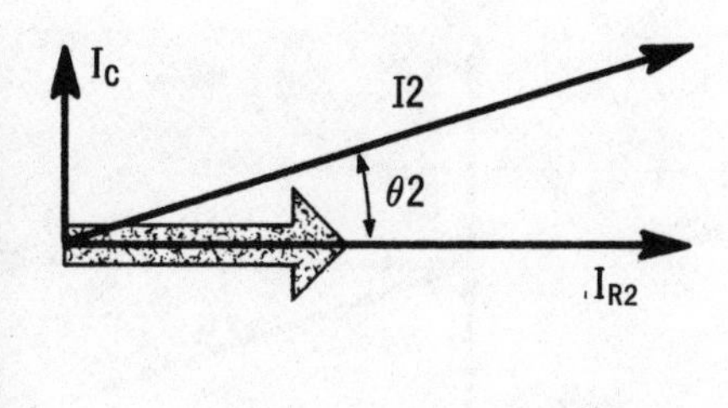

$I_{R2} = I2 \cos \theta 2 = 0.46 \cos 17° = 0.44$ A

$I_C = I2 \sin \theta 2$ (leading) $= 0.46 \sin 17° = 0.13$ A

Series-Parallel Circuits—Solution by Calculation (continued)

The total current is the vector sum of the two branch currents; it can be found by adding the in-phase components ($I_{in\text{-}phase}$) and the out-of-phase components ($I_{out\text{-}of\text{-}phase}$) and then combining them vectorially:

$$I_{in\text{-}phase} = I_{R1} + I_{R2} = 0.29\ A + 0.44\ A = 0.73\ A$$
$$I_{out\text{-}of\text{-}phase} = I_L - I_C \quad \text{or} \quad I_C - I_L = 0.38A - 0.13A = 0.25A\ \text{(lagging)}$$

Although the out-of-phase components are added, remember that they are in opposition; their algebraic sum is therefore the difference between them, and the direction of phase (leading or lagging) is the same as that of the larger one.

$$\text{Total current, } I_t = \sqrt{(I_{R1} + I_{R2})^2 + (I_L - I_C)^2}$$
$$= \sqrt{(0.73)^2 + (0.25)^2}$$
$$= 0.77\ A$$
$$\text{Phase angle } \theta = \tan^{-1}\frac{I_{out\text{-}of\text{-}phase}}{I_{in\text{-}phase}}$$
$$= \tan^{-1}\frac{0.25}{0.73} = 19^\circ$$
$$\text{Total impedance, } Z_t = E/I_t = 240/0.77 = 312\ \text{ohms}$$

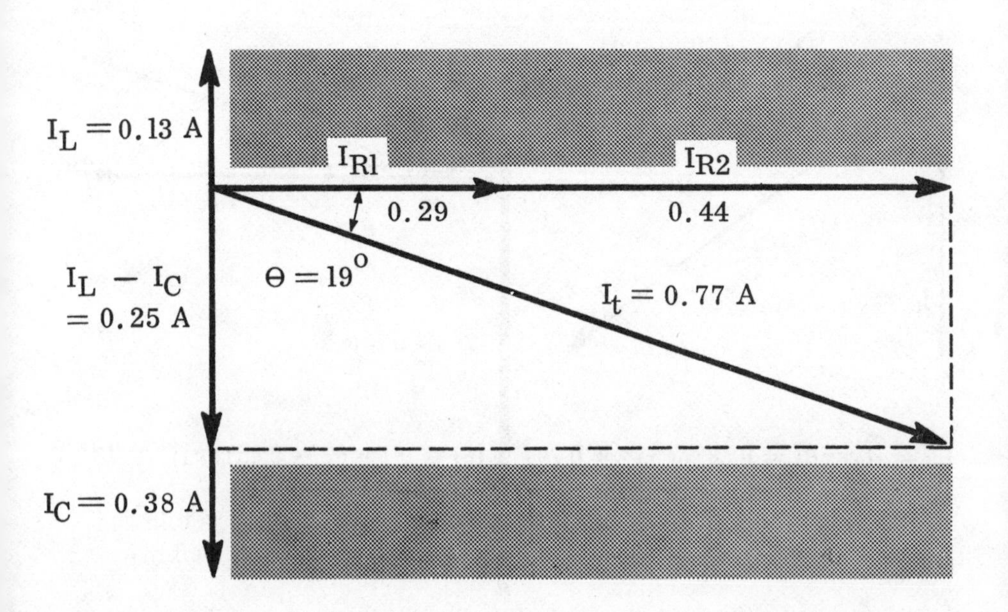

More Complex Series-Parallel Circuits—Vector Graphic Solution

Series-parallel circuits may be even more complex than the one just illustrated. For example, suppose that the series-parallel circuit under discussion is connected in series with an inductance and a resistance.

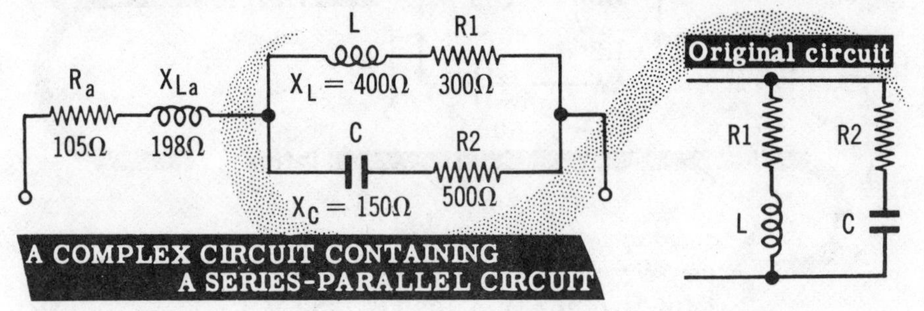

To calculate the impedance of the circuit, first resolve that part of the circuit considered earlier (see pages 4-78 through 4-81). In order to do this, you must assume an arbitrary voltage appearing across it.

This portion of the circuit, therefore, behaves as an inductor L_e and resistor R_e in series, having an effective impedance of 312 ohms and with the voltage leading the current by 19 degrees. Now draw the impedance sketch as shown below.

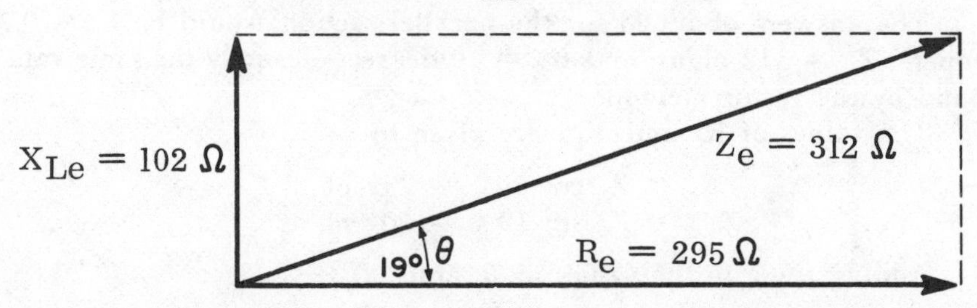

Z_e can be resolved into R_e and X_{Le} by vectors with values of 295 ohms and 102 ohms, respectively. These values are combined with those of R_a and X_{La}, giving resultants of 400 ohms (295 + 105) for R_t and 300 ohms (102 + 198) for X_{Lt} . Now draw the final impedance sketch.

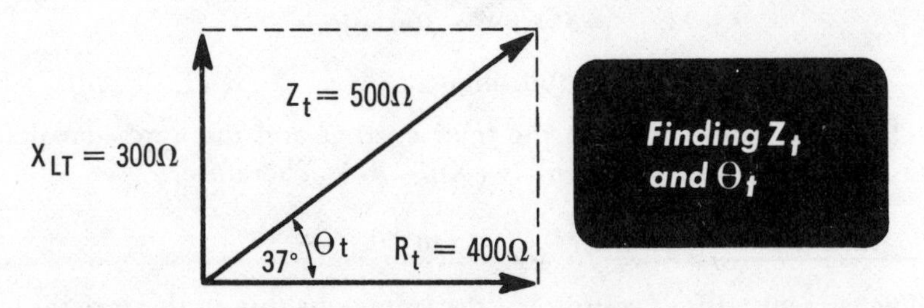

Thus, Z_t is 500 ohms, and the phase angle is 37 degrees, with the current lagging the applied voltage since the circuit is inductive.

More Complex Series-Parallel Circuits—Solution by Calculation

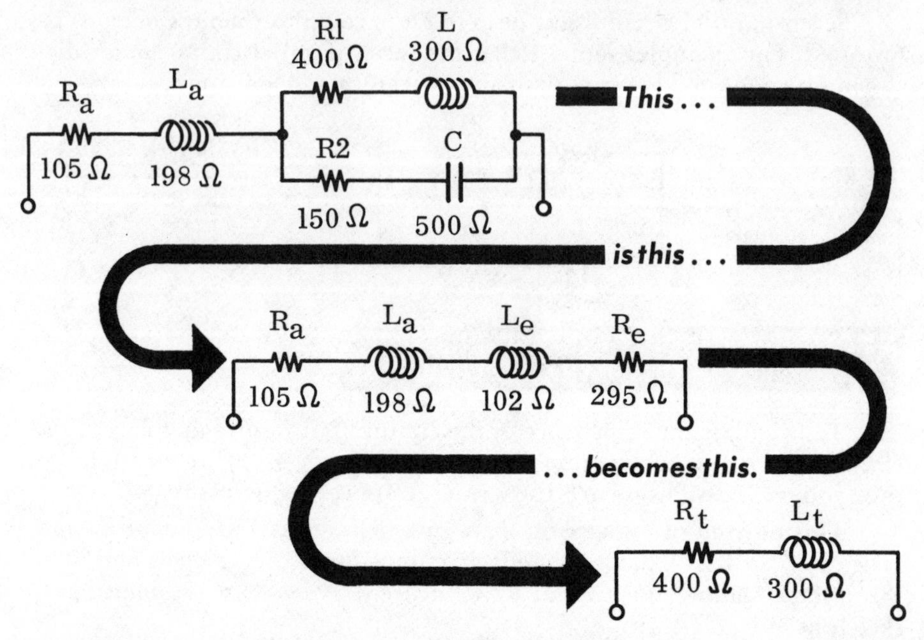

The same problem could be solved mathematically by using the method given on page 4-81.

The answers obtained for the parallel section would be $I_e = 0.77$ ampere, $Z_e = 312$ ohms, and $\theta_e = 19$ degrees—exactly the same values found by the vector method.

The values of R_e and X_{Le} are given by

$$R_e = Z_e \cos 19^\circ = 295 \text{ ohms}$$
$$X_{Le} = Z_e \sin 19^\circ = 102 \text{ ohms}$$

Adding these to the values of R_a and X_{La}, you get

$$R_t = 400 \text{ ohms} \quad \text{and} \quad X_{Lt} = 300 \text{ ohms}$$

Thus

$$Z_t = \sqrt{R_t^{\ 2} + X_{Lt}^{\ 2}} \text{ ohms}$$
$$= \sqrt{400^2 + 300^2} \text{ ohms}$$
$$= 500 \text{ ohms}$$

The phase angle between the total current and the applied voltage (θ_t) is the angle whose tangent is (X_{Lt} / R_t). Therefore,

$$\theta_t = \tan^{-1} \frac{300}{400} = \tan^{-1} 0.75 = 37^\circ$$

The current will, of course, lag the voltage by this angle since the circuit is inductive.

Experiment/Application—Graphic and Calculated Solution in Complex Circuits

In this experiment/application, you will be using both of the methods described on pages 4-78 through 4-81 to find—in a series-parallel circuit containing resistance, inductance, and capacitance—the total circuit current, the branch currents, and the impedance. Your graphic and calculated results will then be checked with *actual* voltage and current measurements, and you will see how those values compare.

Because pure inductances and pure capacitances are only *theoretical* possibilities, there will be a difference between the *measured* and the *calculated* results, but the measured results will show that the calculations are accurate enough for practical use in electric circuits.

Connect a 500-ohm resistor, a 1,000-ohm resistor, a 1-μF capacitor, and a 5-henry inductor to form the complex circuit shown below. Because the inductor has a dc resistance of approximately 50 ohms, the total resistance of the R and L branch of the circuit is 1,050 ohms, and R2 is shown as a 1,050-ohm resistor rather than as a 1,000-ohm resistor. The transformer is a 120-volt to 60-volt step-down transformer, the circuit being connected across the secondary.

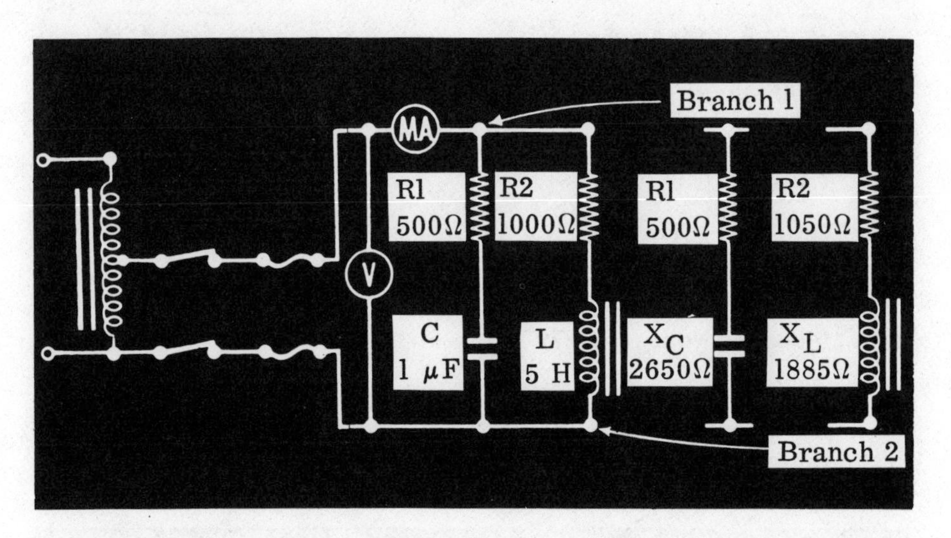

Before switching on, calculate the impedance of, and the current through, the circuit. First the values of X_L and X_C are computed, using 60 Hz as the line frequency. Rounded off to the nearest 10 ohms, these values are 1,885 ohms for X_L and 2.650 ohms for X_C.

Using the known values of R1 and R2 together with the computed values of X_L and X_C, the impedances of each series branch are found separately by using vectors or by calculation. From these values of impedance and a source voltage of 60 volts, the values of I_1 and I_2 are found.

Experiment/Application—Graphic and Calculated Solution in Complex Circuits (continued)

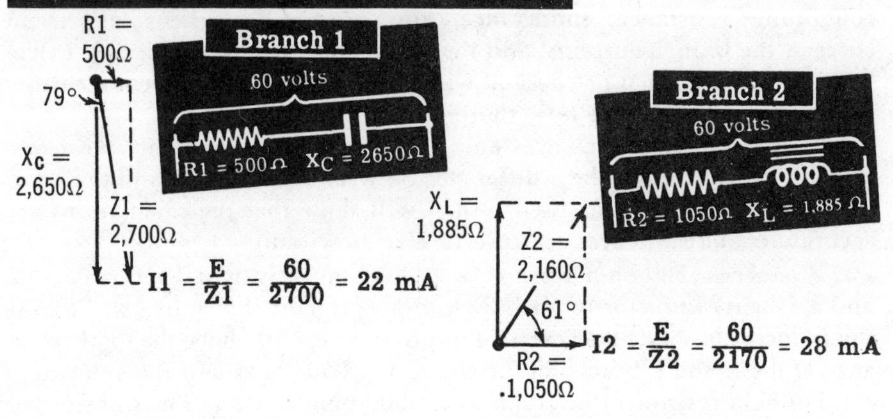

$$Z1 = \sqrt{R1^2 + X_C^2}$$

$$= \sqrt{500^2 + 2{,}650^2}$$

$$= 2{,}700 \text{ ohms approx.}$$

$$I1 = \frac{E}{Z1} = \frac{60}{2{,}700} \cong 22 \text{ mA}$$

$$\theta 1 = \tan^{-1} \frac{X_C}{R1} = \tan^{-1} \frac{2{,}650}{500}$$

$$= \tan^{-1} 5.3 = 79^{\circ}$$

I1 *leads* applied voltage by 79°

$$Z2 = \sqrt{R2^2 + X_L^2}$$

$$= \sqrt{1{,}050^2 + 1{,}885^2}$$

$$= 2{,}160 \text{ ohms approx.}$$

$$I2 = \frac{E}{Z2} = \frac{60}{2{,}160} = 28 \text{ mA}$$

$$\theta 2 = \tan^{-1} \frac{X_L}{R2} = \tan^{-1} \frac{1{,}885}{1{,}050}$$

$$= \tan^{-1} 1.8 = 61^{\circ}$$

I2 *lags* applied voltage by 61°

The branch current vectors are now drawn with reference to the common total voltage vector, E, and the two branch current vectors are combined to find the total circuit current. From this value of the total current and the given value of voltage, 60 volts, the total impedance is computed.

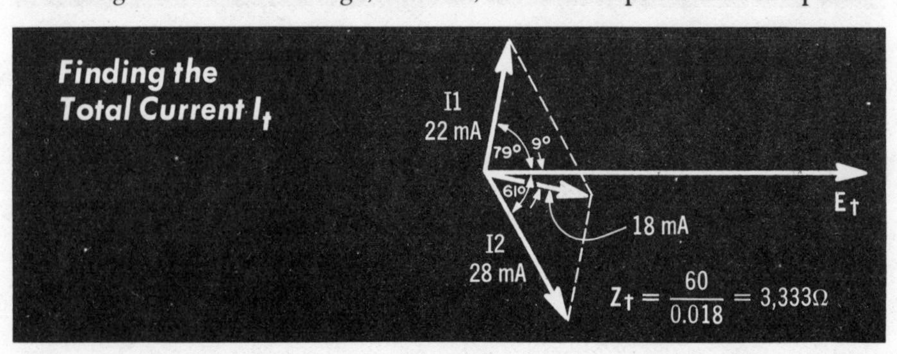

Experiment/Application—Graphic and Calculated Solution in Complex Circuits (continued)

By calculation, you can solve for the total current by resolving each current (I_t and I2) into their respective components, adding these components, and solving for the total current:

$$I_{R1} = I1 \cos \theta 1 = 22 \cos 79^{o} = 4.2 \text{ mA}$$

$$I_{R2} = I2 \cos \theta 2 = 28 \cos 61^{o} = 13.6 \text{ mA}$$

$$I_C = I1 \sin \theta 1 = 22 \sin 79^{o} = 21.6 \text{ mA}$$

$$I_L = I2 \cos \theta 2 = 28 \sin 61^{o} = 24.5 \text{ mA}$$

$$I_{R1} + I_{R2} = 4.2 + 13.6 = 17.8 \text{ mA}$$

$$I_L - I_C = 24.5 - 21.6 = 2.9 \text{ mA}$$ (inductive; therefore total current lags voltage)

$$I_t = \sqrt{(I_{R1} + I_{R2})^2 + (I_L - I_C)^2} = \sqrt{(17.8)^2 + (2.9)^2}$$

$$= 18 \text{ mA} = 0.018 \text{ A}$$

$$Z_t = \frac{60}{0.018} = 3{,}333 \text{ ohms}$$

The θ angle of the current is given by

$$\theta = \tan^{-1} \frac{(I_L - I_C)}{I_R}$$

$$\approx 9^{o}$$

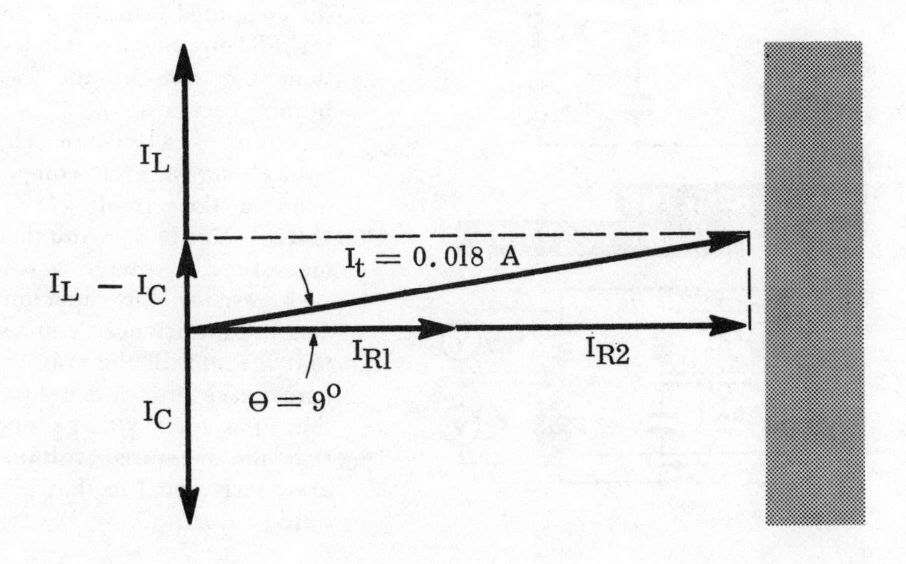

Experiment/Application—Verification of Calculated Current and Impedance

To check the computed values of the total current and circuit impedance, close the switch and measure the total circuit current and voltage. You will see that the measured total current is approximately the same as the computed value. From these measured values, the impedance Z_t is determined and compared with the value obtained by calculation.

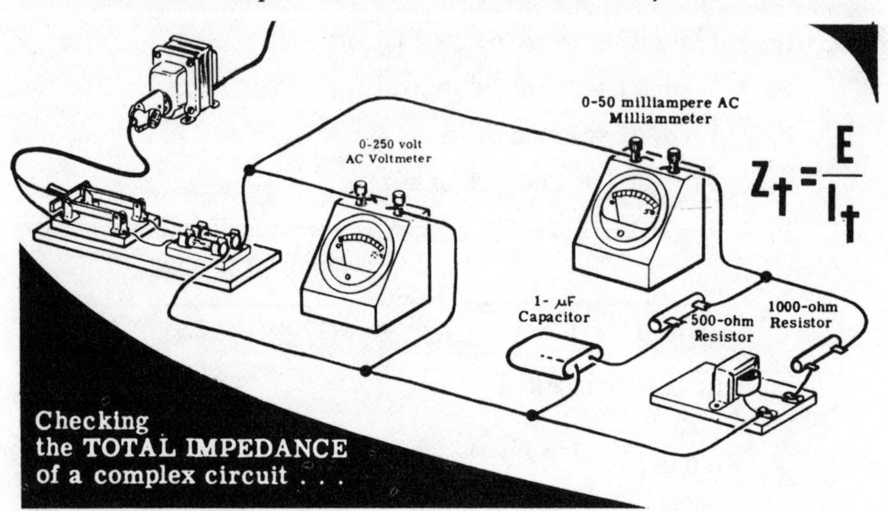

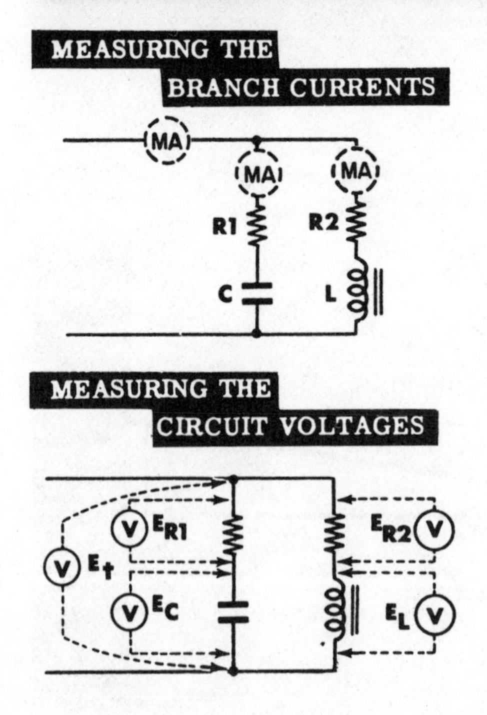

Next connect the milliammeter to measure each branch current in turn. You see that the measured values are about the same as the computed values and that the total circuit current is less than the sum of the two branch currents.

Now calculate the voltage across each component in the circuit (I_1R1, I_2R2, I_1X_C, I_2X_L), and then measure the voltage across each resistor, the capacitor, and the inductance. You see that the sum of the voltages across each branch is greater than the total voltage and that the measured voltages are nearly equal to the computed values.

Review—AC Complex Circuits

The solution of an ac complex circuit requires a step-by-step use of vectors and Ohm's law to find unknown quantities of voltage, current, and impedance. Suppose you review the vector solution of a typical complex circuit either graphically or by calculation.

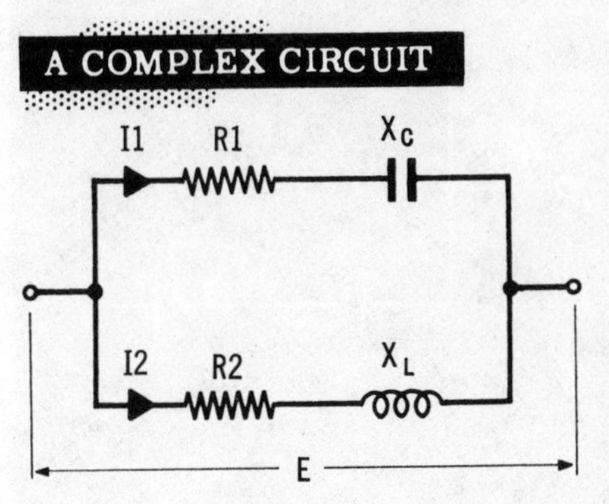

1. Calculate the reactance values of the circuit capacitances and/or inductances.

2. Using vectors, or by calculation, find the impedance of each series branch separately, and compute the Ohm's law values of the branch currents.

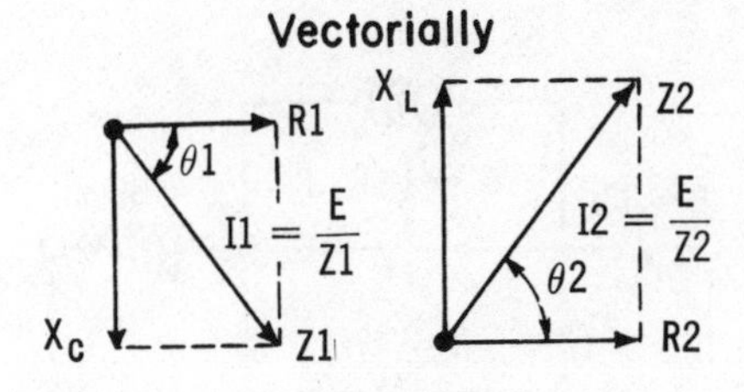

3. You can find the individual voltages of each branch, with respect to their particular branch current. Complete the voltage parallelograms to find the phase relationship between each branch current and the total circuit voltage, or use the formulas for calculation.

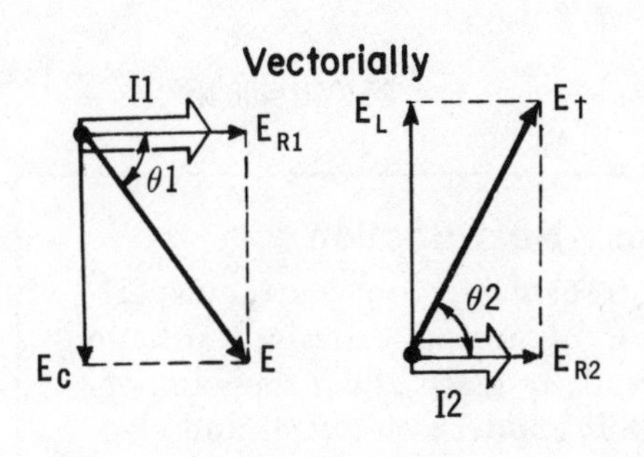

4. Now find the branch current vectors with respect to a common total circuit voltage vector; then combine the currents to find the total circuit current. Using this total current, you can find the total circuit impedance.

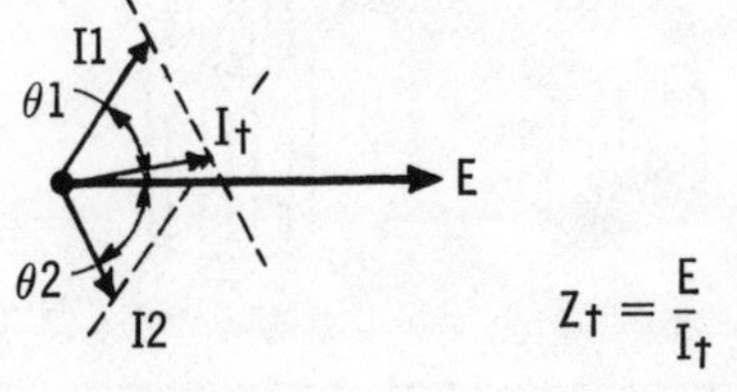

$$Z_t = \frac{E}{I_t}$$

5. You can now calculate the true power, apparent power, and power factor.

Self-Test—Review Questions

1. For the circuits shown below, determine graphically and by calculation the line current and its phase. Also find the true power, apparent power, and power factor.

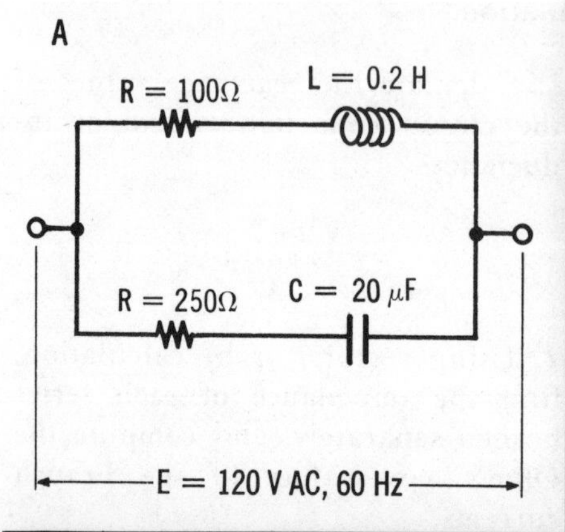

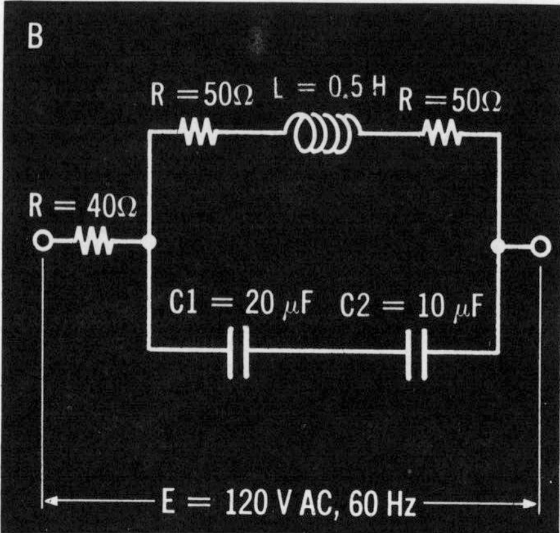

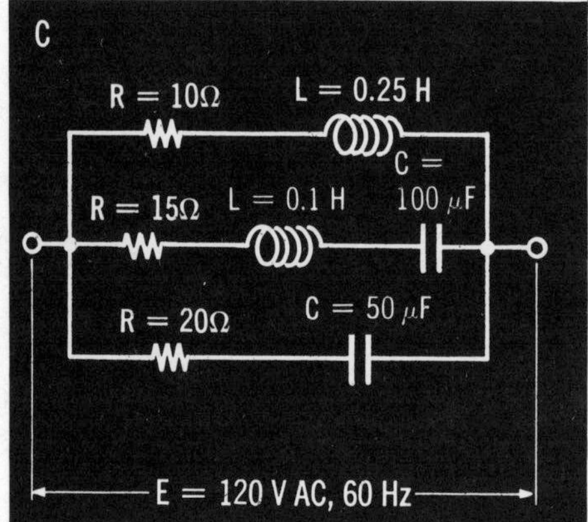

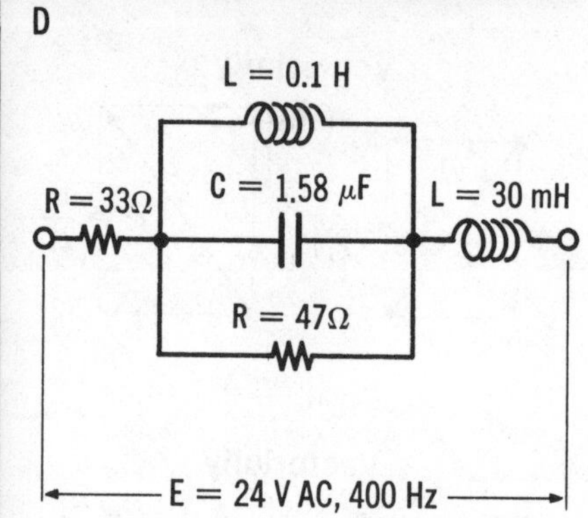

Learning Objectives—Next Section

Overview—You have been using transformers in your ac experiments/applications as a means for obtaining a desired voltage from a line source. You are now ready to learn about these important devices that are indispensable in modern electrical and electronic circuits and equipment.

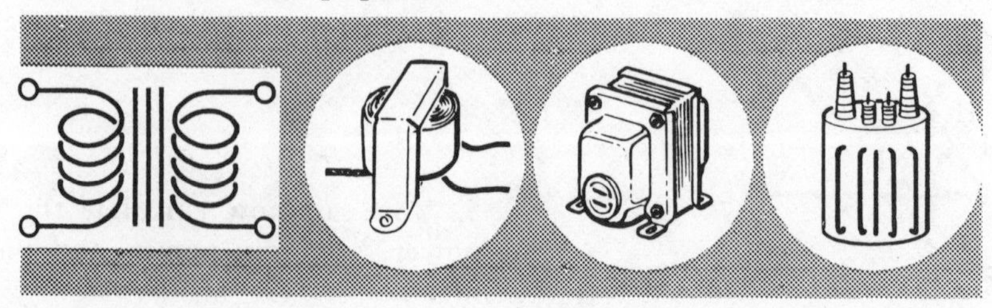

Mutual Induction—Faraday's Law

As you know, ac current is used because it can be readily stepped (transformed) upward or downward in voltage by means of *transformers.* You will now learn that transformers are inductances *coupled* together by their mutual magnetic fields, or *mutual induction.* To start our study of transformers, we will need to learn about mutual induction.

Michael Faraday, an English scientist, did a great deal of important work in the field of electromagnetism, and his work on mutual induction eventually led to the development of the transformer. Faraday found that if the total magnetic flux linking a circuit changes with time, an emf is induced in the circuit. He also found that if the rate of flux change is increased, the magnitude of the induced emf is increased as well. Stated in other terms, Faraday found that the character of an emf induced in a circuit depends upon (a) the amount of flux, and (b) the rate of change of flux which links a circuit. These effects are what is meant by mutual induction.

You have seen in Volume 1 an illustration of the mutual induction principle just stated. When a conductor is made to move with respect to a magnetic field, an emf is induced in the conductor which is directly proportional to the velocity of the conductor's movement with respect to the field. Moreover, the voltage induced in a coil is proportional to the number of turns of the coil, the magnitude of the inducing flux, and the rate of change of this flux.

An example of mutual induction (inducing an emf in a neighboring conductor or inductor) is shown below. Current flows in the direction indicated in primary coil A. This current produces a magnetic field, and if the current remains constant, the number of flux lines produced is fixed. If, however, the current is reduced by opening the switch, the number of flux lines in coil A is decreased, and consequently the flux linking secondary coil B is decreased also. This changing flux induces an emf in coil B, and a current I_B flows, as shown by the movement of the indicator pointer. Thus, it is seen that energy can be transferred from one circuit to another by the principle of *electromagnetic induction*—mutual induction.

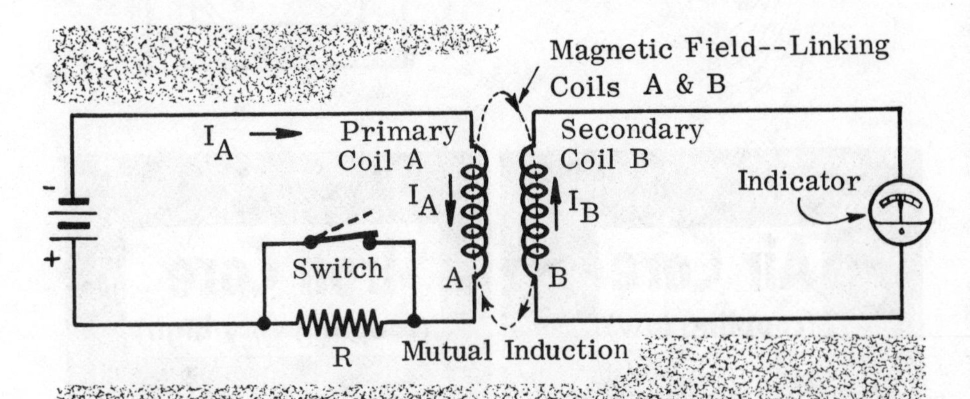

Mutual Induction

Mutual Induction—Faraday's Law (continued)

A battery was used in the experiment on the previous page as a source of emf. The only way current, and thus magnetic field, variations can be developed is by opening or closing the switch. If an ac voltage source with an extremely low frequency (1 Hz) is used to replace the battery, the indicator will show continuous variations in current. The indicator needle moves to the left (or right) first, and then reverses its position, thus showing the reversal in ac current flow.

If the battery from the circuit on the previous page were replaced with 60-Hz ac source in the *primary* coil and the indicator meter with an ac voltmeter in the *secondary* coil, you would find that an ac voltage existed across the secondary terminals. It is obvious that the mutual inductance or coupling between two coils depends on their *flux linkages*. Maximum coupling is when all the flux lines from the primary cut the secondary winding. When this happens, the degree of coupling (called the *coefficient of coupling*) is unity. While coupling factors of less than unity are desirable in some electronic circuits you may learn about later, in transformers it is desirable to get the coupling between the two windings as *high* as possible. Thus, both coils are usually wound on an iron core so that the path of the flux lines can be controlled and kept where desired. A special alloy of silicon steel is commonly used for transformer cores.

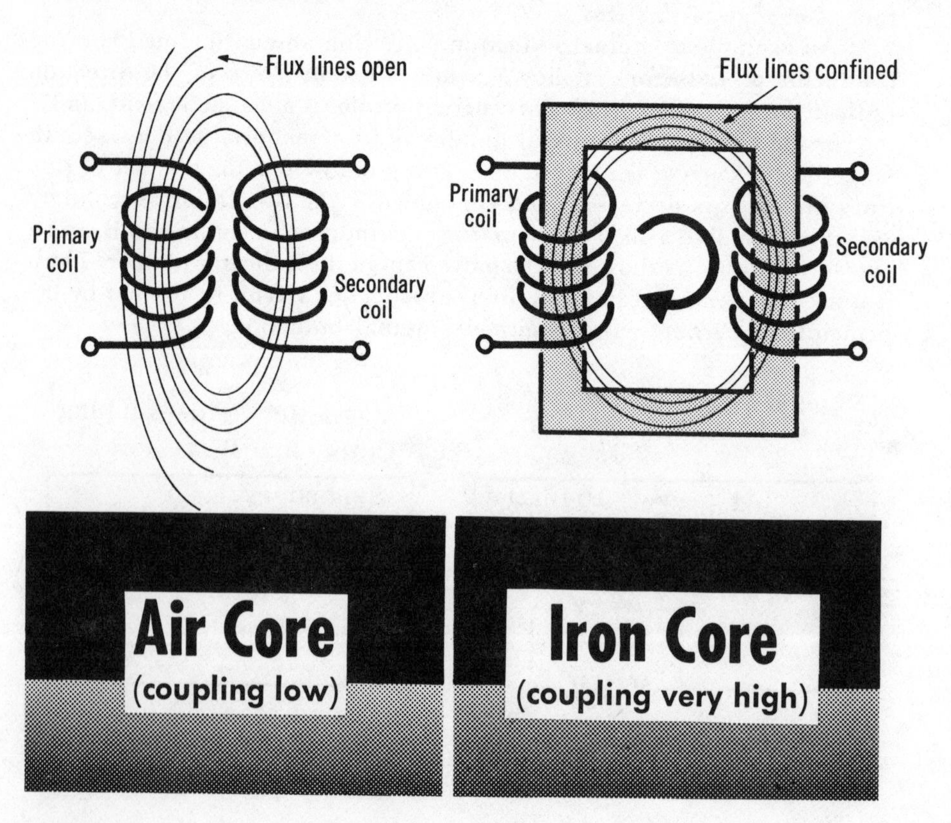

How a Transformer Works

A simple transformer consists of two windings very tightly coupled together, usually with an iron core, but electrically insulated from each other. The winding to which an ac voltage source is applied is called the *primary*. It generates a magnetic field which cuts through the turns of the other coil, called the *secondary*, and generates a voltage in it. The windings are not physically connected to each other. They are, however, *magnetically coupled* to each other. Thus, a transformer transfers electrical power from one coil to another by means of an *alternating magnetic field.*

Assuming that all the magnetic lines of force from the primary cut through all the turns of the secondary, the voltage induced in the secondary will depend on the ratio of the number of turns in the secondary to the number of turns in the primary. This is mathematically expressed as $E_s = (N_s/N_P)\ E_P$. For example, if there are 1,000 turns in the secondary and only 100 turns in the primary, the voltage induced in the secondary will be *10 times* the voltage applied to the primary (1,000/100 = 10). Since there are more turns in the secondary than there are in the primary, the transformer is called a *step-up transformer.* If, on the other hand, the secondary has 10 turns and the primary has 100 turns, the voltage induced in the secondary will be *one-tenth* of the voltage applied to the primary (10/100 = 1/10). Since there are less turns in the secondary than there are in the primary, this transformer is called a *step-down transformer.* The symbol for a transformer is shown below.

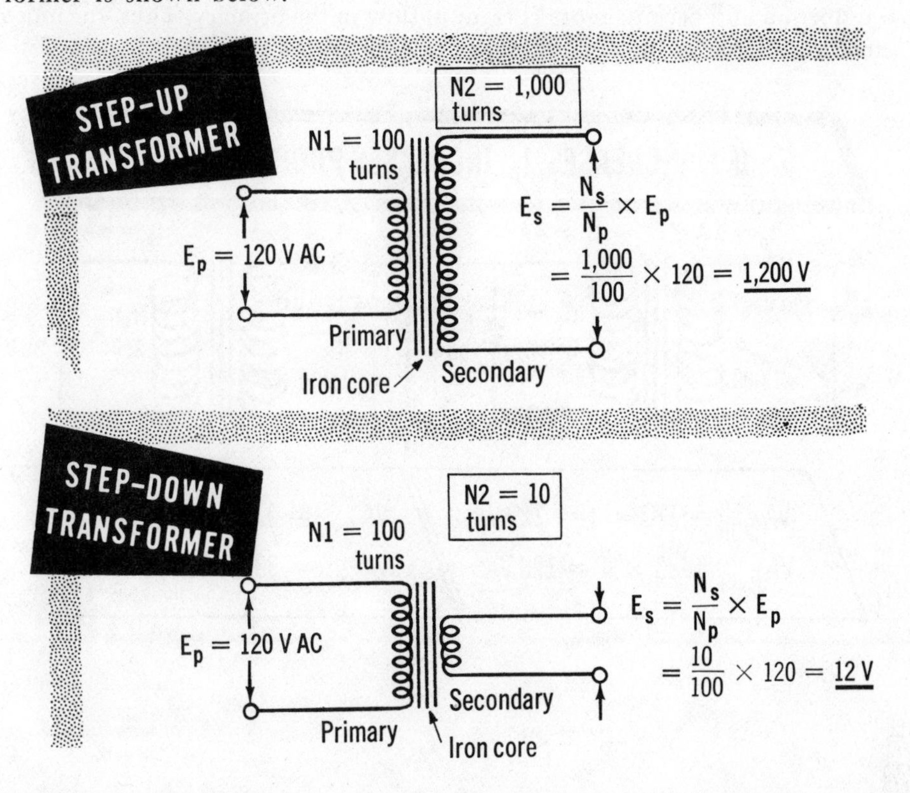

How a Transformer Works (continued)

To find any unknowns in a transformer, use the formula $E_P/E_S = I_S/I_P = N_P/N_S$ and cross-multiply to find the required information.

The current in the secondary of a transformer flows in a direction opposite to that which flows in the primary because of the emf of mutual induction. An emf of self-induction is also set up in the primary which is in opposition to the applied emf.

When no load is present at the output of the secondary, the primary current is very small because the emf of self-induction is almost as large as the applied emf. If no load is present at the secondary, there is no secondary current, but there is a small primary current flow to account for transformer losses. Since there are no fields caused by secondary current flow, the magnetic field of the primary will develop to its maximum strength. When the primary field is at its maximum strength, it produces the strongest possible emf of self-induction, and this opposes the applied voltage. The difference between the emf of self-induction and the applied emf causes a small current to flow in the primary, and this is the exciting or magnetizing current that accounts for transformer losses.

Since any current made to flow in the secondary is opposite to the current in the primary, the lines of flux are opposite, too. As a load is applied to the secondary, causing current to flow, it causes a reduction in the total flux (primary flux–secondary flux) which reduces the flux linking the primary. The reduction in flux lines reduces the emf of self-induction and permits more current to flow in the primary. Thus, the more current in the secondary, the more current in the primary.

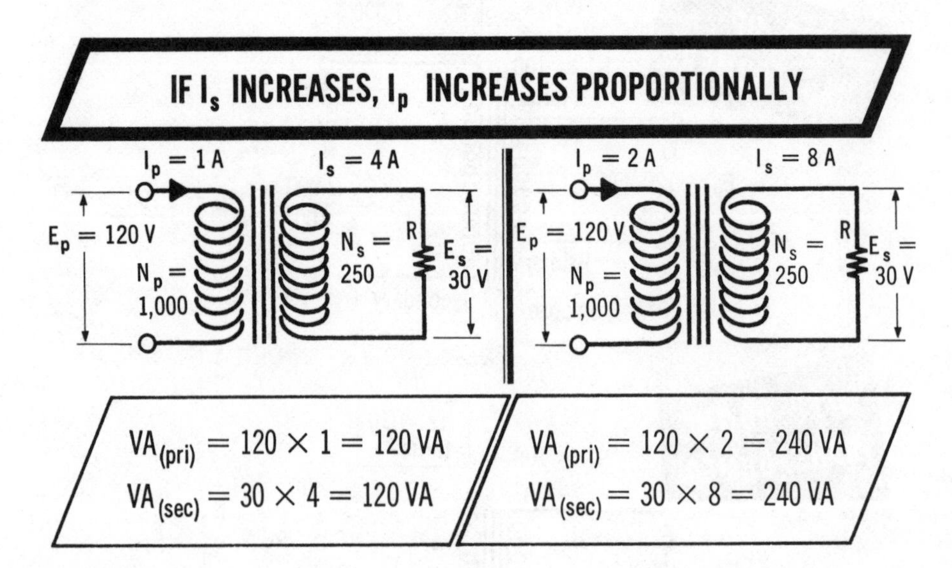

How a Transformer Works (continued)

A transformer *does not* generate electrical power. It simply *transfers* electrical power from one coil to another by magnetic induction. Although transformers are not 100 percent efficient (lossless), they are very nearly so. (You will find out about some of the losses in transformers a little bit later.) Therefore, a transformer can be defined as *a device that transfers power from its primary circuit to its secondary circuit with little loss.*

Transformers are usually rated in volt-amperes rather than watts since they must handle the total current no matter what the phase is. Since power equals voltage times current, if $E_p I_p$ represents the primary power and $E_s I_s$ represents the secondary power, then $E_p I_p = E_s I_s$. If the primary and secondary voltages are equal, the primary and secondary currents must also be equal.

Suppose E_p is twice as large as E_s. Then, for $E_p I_p$ to equal $E_s I_s$, I_p must be one half of I_s. It follows that a transformer which *steps voltage down* must *step current up.*

Similarly, if E_p is only half as large as E_s, I_p must be twice as large as I_s. So a transformer which *steps voltage up must step current down.*

Transformers are classified as step-down or step-up by reference to their effect on *voltage only.*

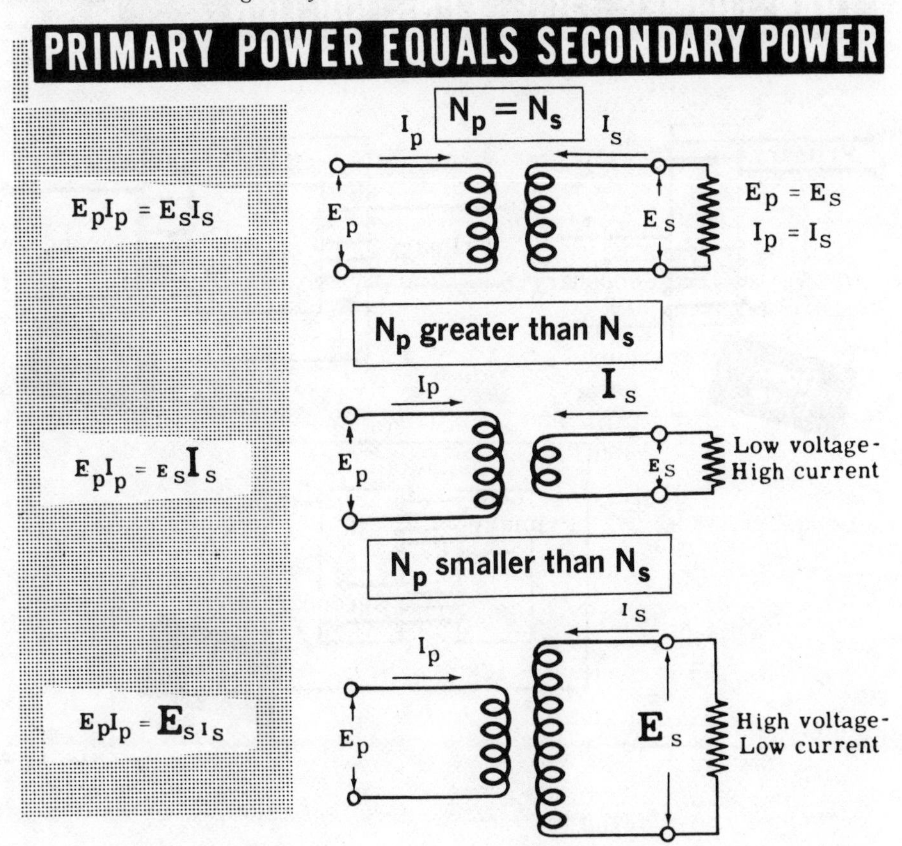

Transformer Construction

Transformers designed to operate on low frequencies (power transformers) have their coils, called *windings*, wound on iron cores. Since iron offers little resistance to magnetic lines, nearly all the magnetic field of the primary flows through the iron core and cuts the secondary.

Iron cores are constructed in three main types—the open core, the closed core, and the shell core. The open core is the least expensive to build since the primary and the secondary are wound on one cylindrical core. The magnetic path is partly through the core, partly through the air. Since the air path opposes the magnetic field, the magnetic interaction or *linkage* is weakened. The open core transformer is therefore inefficient and is never used for power transmission.

The closed core improves the transformer efficiency by offering more iron paths and less air path for the magnetic field, thus increasing the magnetic linkage or *coupling*. The shell core further increases the magnetic coupling, and therefore the transformer efficiency, because it provides two parallel magnetic paths for the magnetic field. It thus permits maximum coupling to be attained between the primary and the secondary.

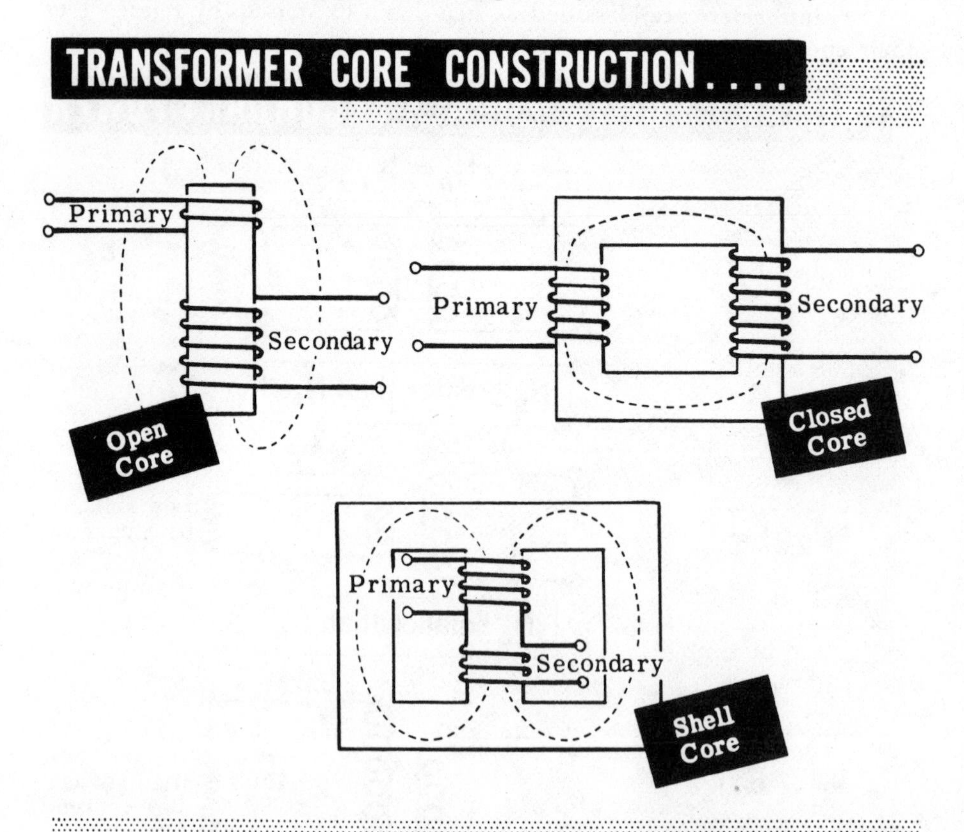

Transformer Losses

Up to now, you have assumed that transformers are perfect and have no internal losses. While this is usually almost true, there are some losses present. Most transformers are between 90 and 99 percent efficient. The major transformer losses are copper loss (resistance loss), flux leakage loss, hysteresis loss, eddy current loss, and saturation loss.

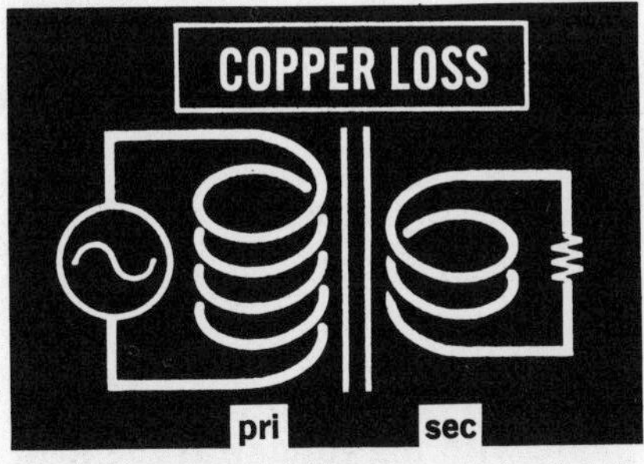

Loss from resistance of (copper) wire in windings of transformer = I^2R, where R is winding resistance.

Loss from flux lines that leak from windings or core so that they do not link the flux of primary and secondary and, therefore, represent lost energy.

HYSTERESIS LOSS

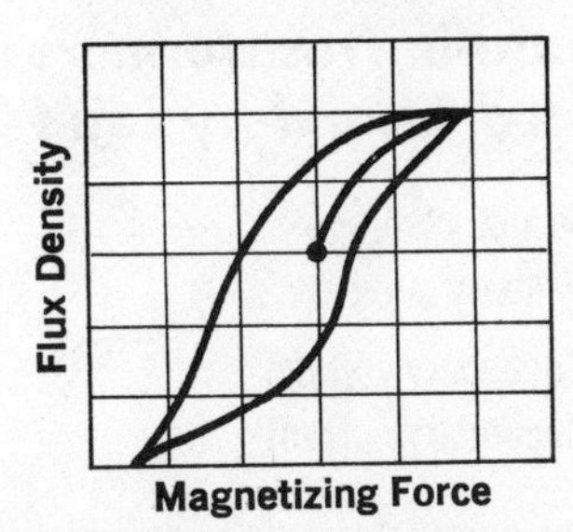

When the current reverses, magnetic alignment of the core also reverses, but the magnetic domains are a little behind. Energy is needed to get the magnetic domains aligned and to turn them around. This energy, which is not available in the secondary, is called **hysteresis loss.** Some transformers employ a powdered iron core to reduce these losses.

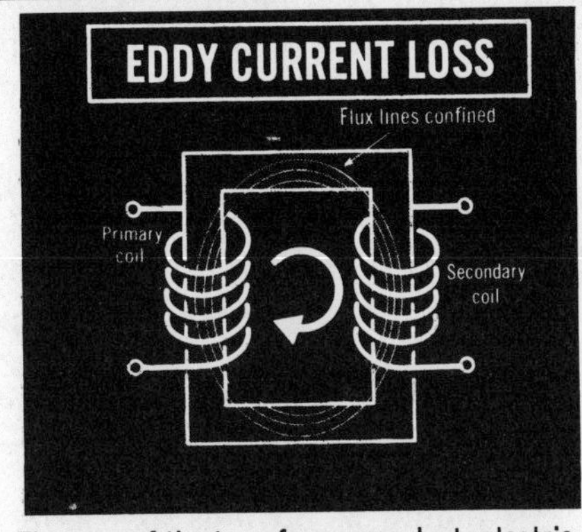

The core of the transformer conducts electricity and acts like a single-turn shorted secondary. The current that flows in the core is called **eddy current.** Eddy currents are kept at a minimum by using a core made up of many flat sections, called **laminations,** stacked together.

SATURATION LOSS

As primary current increases, the number of flux lines in the core increases. Finally, a limit is reached, and additional current results in no more flux lines. When this point is reached, the core is said to be **saturated,** and to increase the number of flux lines would require that the core be made larger.

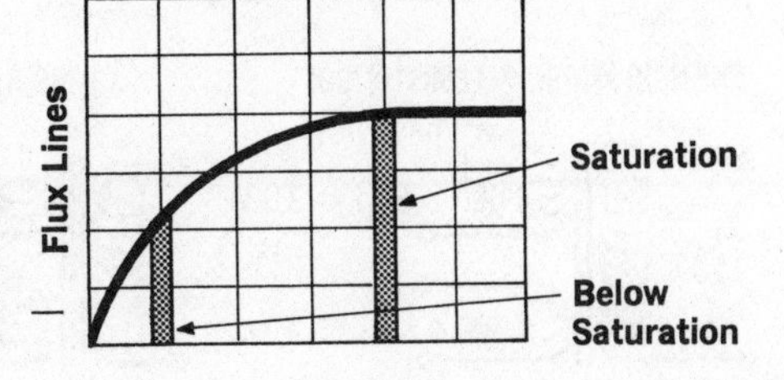

Phase Relationships in Transformers—Multiple Winding Transformers

A transformer with no load on the secondary (open circuit) acts like a simple inductor. Since the losses are usually small, as you might suspect, the current lags the voltage by 90 degrees. The secondary voltage is 180 degrees out of phase with the applied voltage.

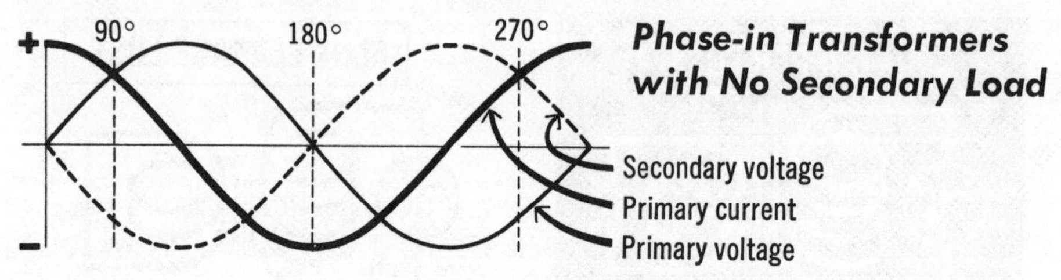

Phase-in Transformers with No Secondary Load

If we put a variable load resistance on the secondary, we would find that the current moves toward being in phase with the voltage as the load resistance increases. When the load current increases so that it is much larger than the no-load current (the way transformers normally operate), the voltage and current in the primary and the secondary have the same relative phase relationship, even though their magnitudes may be different. Thus, a load with a less than unity power factor on the secondary will be reflected as a load with the *same* power factor in the primary.

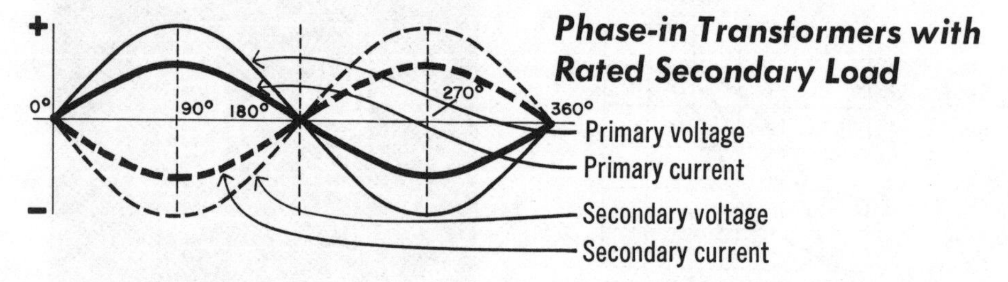

Phase-in Transformers with Rated Secondary Load

It should be obvious that more than one secondary can be put on a transformer. When this is done, the phase relationship between all secondaries will be the same if the loads are similar. Sometimes, it is desirable to connect secondary windings in series so that the voltage will be increased or decreased. When the windings are connected in *series aiding*, the voltage is the *sum* of the voltages for both windings. When the windings are connected in *series opposing*, the voltage is the *difference* of the voltages for both windings. Sometimes, a dot or similar marking is used to indicate the terminals that have the same phase.

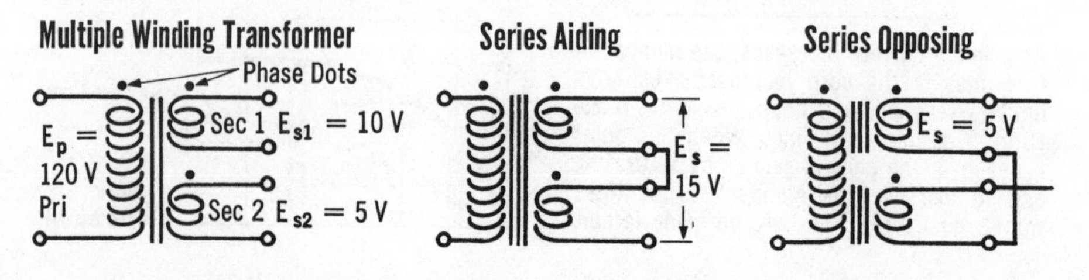

Autotransformers—Tapped Transformers

Autotransformers make common use of part of a winding for both the primary and secondary. They have a tap (a wire brought out from a point on the winding) that is necessary for their operation. As shown below, the operation of an autotransformer is exactly like that of a conventional transformer where one lead of the primary and one of the secondary are tied together in the right way.

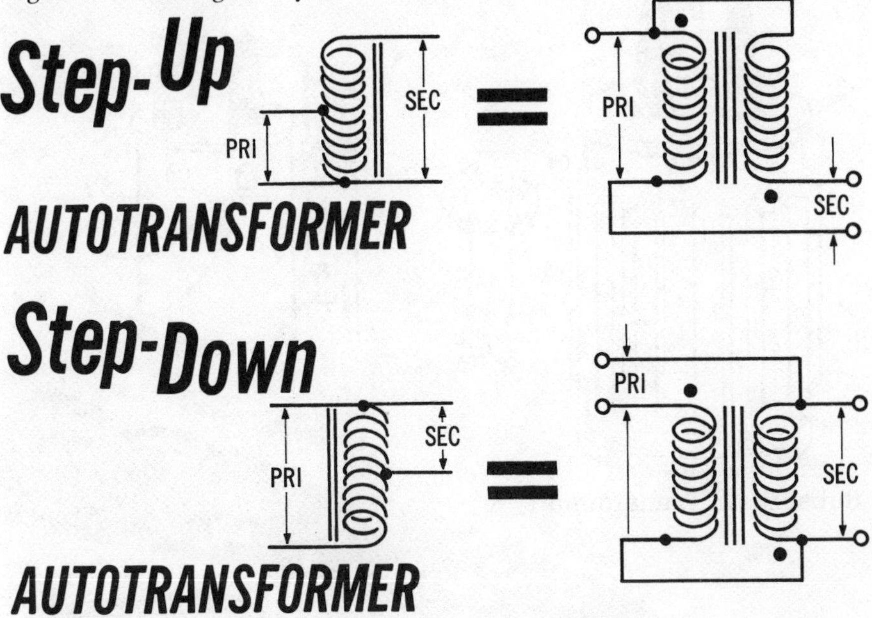

A common type of autotransformer is arranged so that the tap is continuously adjustable in order to provide a voltage range from 0 to about 130 percent of the nominal voltage available. The variable transformer is useful in circuits where it is necessary to set the voltage precisely to a known value.

In some cases, when continuous adjustment is not necessary, fixed taps are used to change the turns ratio. In such cases, the taps can be used in an autotransformer configuration or as taps on the primary or secondary of a conventional transformer.

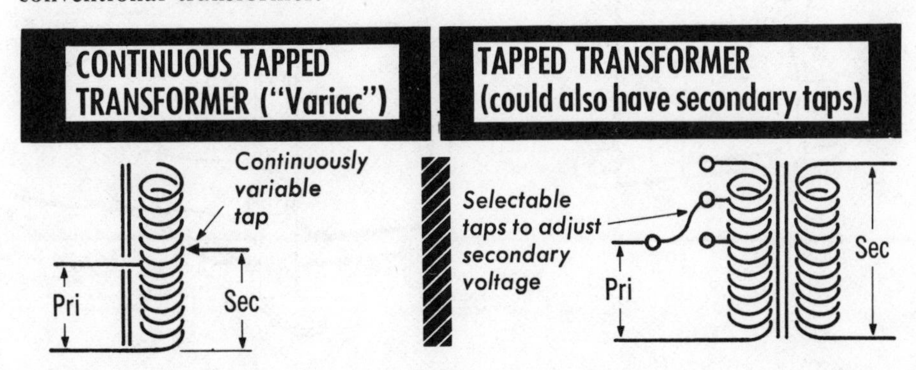

Transformer Types

Transformers come in all sizes and shapes. Very large, high-voltage, high-current transformers are used for power distribution; subminiature transformers are used in highly portable, low-power equipment. Large transformers are sometimes immersed in oil for cooling and improved insulation.

Experiment/Application—Transformer Action

You can observe a transformer in action by making a very simple one and seeing how it works. Suppose you had an iron bar (a bundle of soft iron wires would be better) and you wound a layer of thin insulating tape on this. You could then take some No. 22 or No. 24 enameled wire and wind three coils of it as follows: Winding 1 = 500 turns, Winding 2 = 100 turns, Winding 3 = 250 turns.

Suppose you then connected up the transformer, using Winding 1 as the primary and Windings 2 and 3 as the secondary, with a 7.5-volt ac source. Using a voltmeter to measure the voltages, you will see that they essentially agree with basic transformer theory.

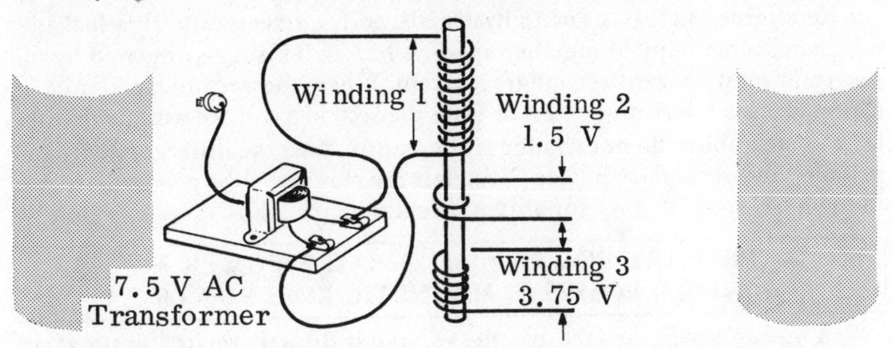

Reverse the setup by using Winding 3 as the primary, and measure the voltages across Winding 1 and Winding 2.

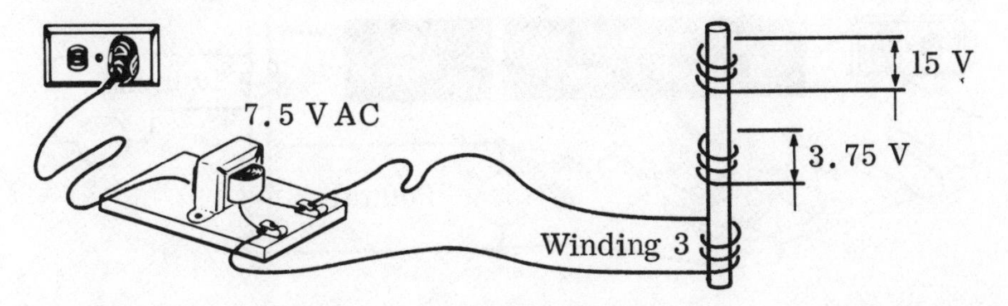

If you have an ammeter available, measure the primary current with no load and calculate the transformer losses. Since this transformer is not very efficient, it will not transfer enough power for you to be able to test the power relationship.

Experiment/Application—Transformer Action (continued)

To demonstrate further how transformers work, you could also take a filament transformer (120 volts to 6.3 volts) and connect it to the line as shown below. If a resistive load is put on the secondary, and primary power and secondary power are measured, the efficiency of the transformer can be found.

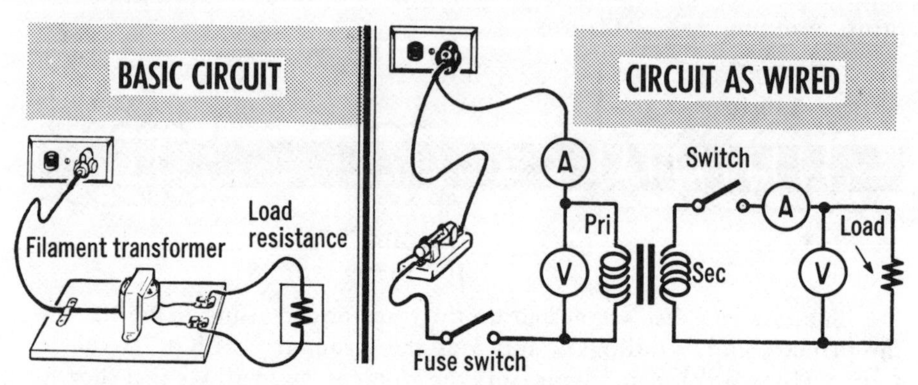

With the primary switch closed and the secondary switch open, you can determine the losses due to hysteresis, eddy currents, and flux leakage, which are often lumped together as *magnetic loss*. This is represented by the power (VA) with zero secondary current. When the secondary switch is closed, the load will draw current from the secondary. If we assume that the losses listed above do not change significantly, then the difference between primary and secondary power (VA), less the magnetic loss power, will give the copper, or I^2R, loss for this particular load. Thus,

PRIMARY POWER = SECONDARY POWER + COPPER LOSSES + MAGNETIC LOSS POWER

You can verify the fact that the voltage is directly related to the turns ratio by using an open winding transformer with a known turns ratio and operating at low voltage.

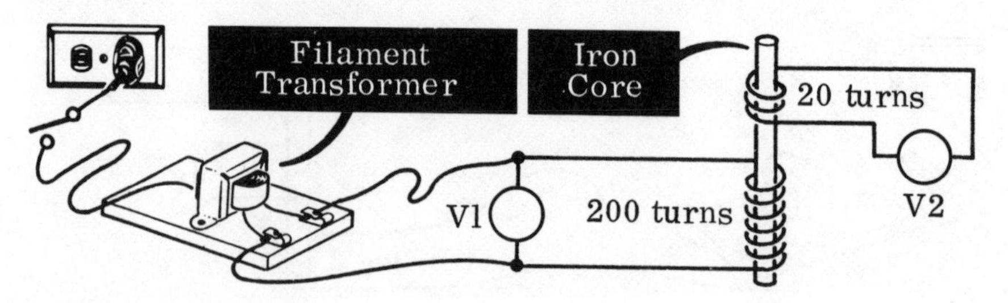

You will see that V1 equals about 6 volts and that V2 is one-tenth of V1, or about 0.6 volt.

$$V2 = \frac{V1 \times N2}{N1} = \frac{6 \text{ volts} \times 20 \text{ turns}}{200 \text{ turns}} = 0.6 \text{ volt}$$

Review of Transformers

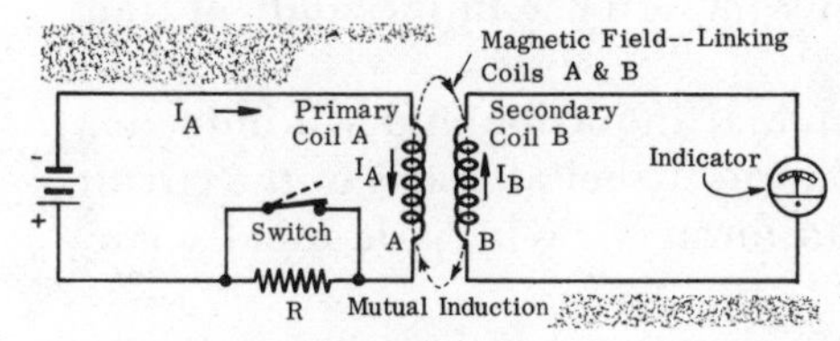

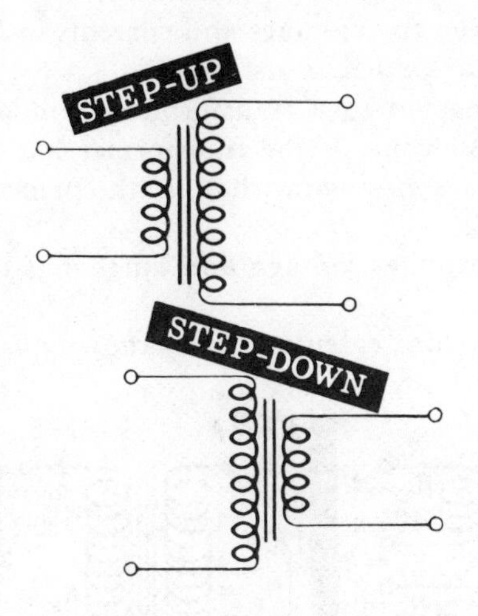

$$E_pI_p = E_sI_s$$

SATURATION LOSS

COPPER LOSS

HYSTERESIS LOSS

EDDY CURRENT LOSS

FLUX LEAKAGE LOSS

Phase-in Transformers with Rated Secondary Load

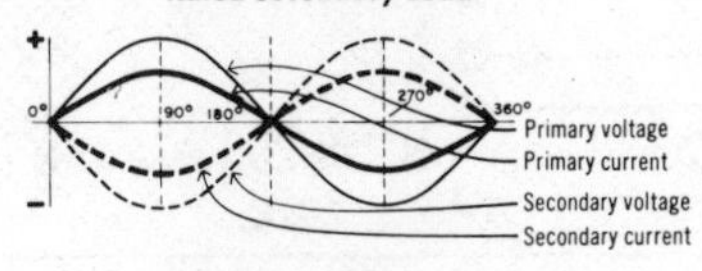

TAPPED TRANSFORMER (could also have secondary taps)

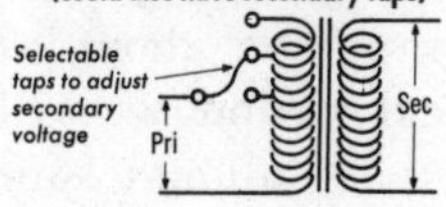

1. MUTUAL INDUCTION—Mutual induction is the induction of an emf in one coil as a result of a current change in a neighboring coil or conductor.

2. TRANSFORMER—A transformer consists of two or more windings coupled together, usually by an iron core, so that almost all flux lines interact with the windings.

3. STEP-UP TRANSFORMER/ STEP-DOWN TRANSFORMER—The voltage ratio in a transformer is proportional to the turns ratio. When the number of secondary turns is *greater* than the number of primary turns, the transformer is a *step-up* transformer. When the number of secondary turns is *less* than the number of primary turns, the transformer is a *step-down* transformer.

4. PRIMARY POWER = SECONDARY POWER—As the turns ratio of a transformer changes, the voltage and current change to keep the primary power equal to the secondary power.

5. TRANSFORMER LOSSES—The losses in transformers, which are usually small, include copper loss, flux leakage loss, hysteresis loss, eddy current loss, and saturation loss.

6. PHASE RELATIONSHIP—In a resistively loaded transformer, the primary voltage and current are essentially in phase, and the secondary voltage and current are also in phase but opposite in polarity to the primary.

7. ADJUSTABLE-VARIABLE TRANSFORMERS—These allow for the adjustment of the turns ratio to permit voltage changes.

Self-Test—Review Questions

1. State Faraday's law and explain its importance in the study of transformers.
2. What is the difference between mutual induction and self-induction?
3. Draw a transformer in schematic form. Label all parts of the circuit.
4. Why is an iron core used in transformers? Why is it usually laminated?
5. List and briefly describe the losses that occur in transformers.
6. Draw the phase relationships between the voltages and currents in (a) an unloaded transformer, and (b) a loaded transformer.
7. You have a load of 50 ohms connected to a transformer secondary which produces an ac voltage of 30 volts. If the transformer has no losses and has a turns ratio of 4-to-1 step-down, what are the primary voltage and current?
8. For the same load, calculate the primary voltage and current if the transformer is a 3-to-1 step-up.
9. For the transformer circuits shown below, calculate the unknown quantities indicated.

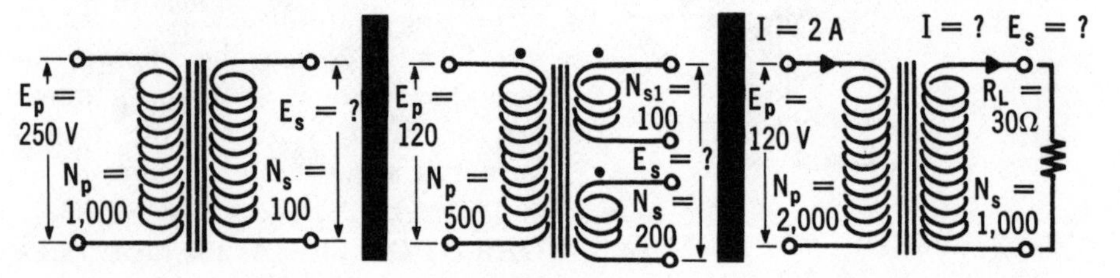

10. Show how the transformer below can be used to obtain the following voltages: 55, 165, 22, 77, and 98 volts.

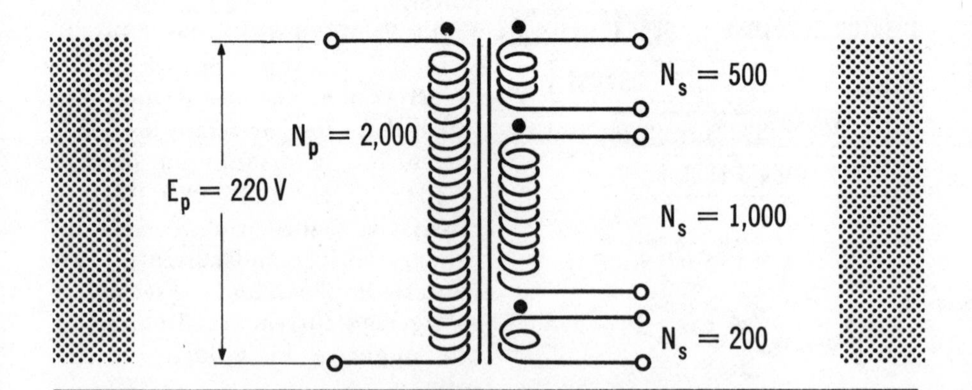

Learning Objectives—Next Section

Overview—One of the most important things to know about is how to troubleshoot ac and dc circuits. In this next section, we will expand on the troubleshooting you have learned in Volumes 2 and 3.

AC Power Systems

As you know, in the U.S. and most North and Central American countries, the power line frequency is 60 Hz and the nominal line voltages for household and industrial use are specified as 120/240 single-phase or 120/208 three-phase. Some older houses have only 120-volt ac single-phase. As you will learn in Volume 5, there are many advantages to the use of three-phase power for some industrial applications and for power transmission. For most homes and very small industrial plants, a single-phase 120/240-volt system is used. This is obtained from the main line that is at a voltage between 600 and 1,000 volts and stepped down by a transformer.

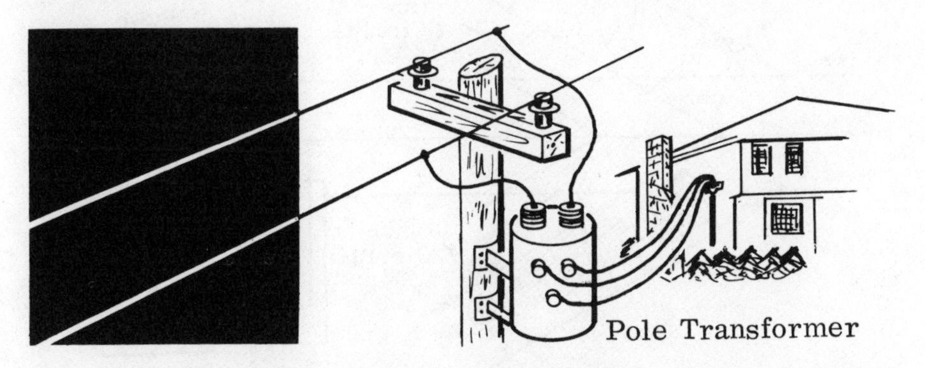

The transformer has a center-tapped secondary so that from each side of the center tap—often called the *neutral*—the line voltage is 120 volts and the total voltage is 240 volts.

Schematically, the system looks as shown below:

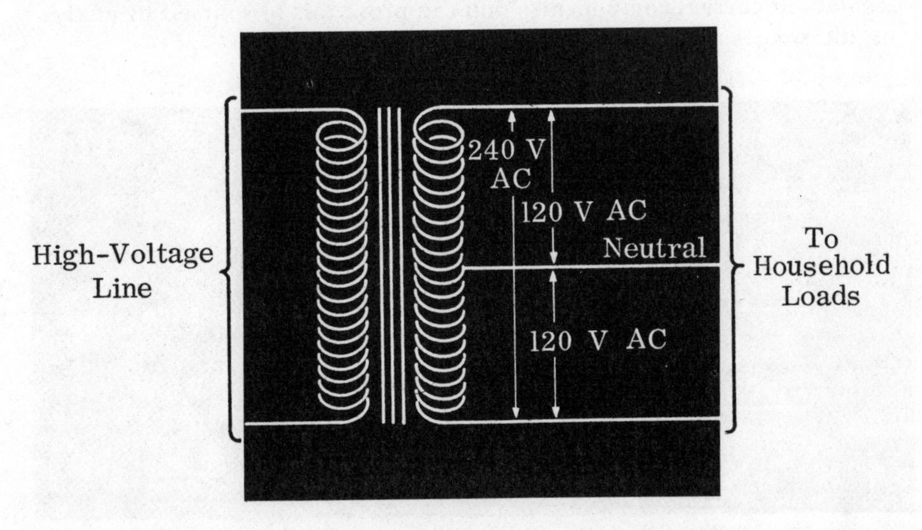

The neutral line is grounded at the pole and at each subscriber's power input box.

Local AC Power Distribution

Usually, the line goes through a recording wattmeter to record the amount of electricity used. How this meter works will be discussed in Volume 5. The power at the household level is distributed via protective devices to the household circuits. For most appliances and lighting, etc., the circuit consists of one side and the neutral so the line voltage is 120 volts ac. For heavy loads (stoves, heaters, dryers, etc.) the appliance is put across the 240-volt lines to reduce the current required.

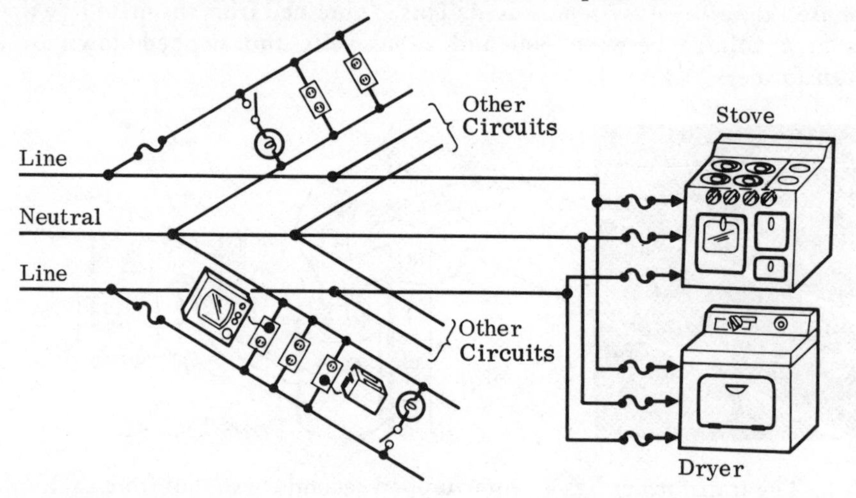

As you can see, the 120-volt circuits are normally loaded evenly so that the line currents are fairly well balanced. If the loads are perfectly balanced there is no current in the neutral. Thus, the neutral only carries the unbalanced current components. You can prove this to yourself by analyzing the simple diagram shown below:

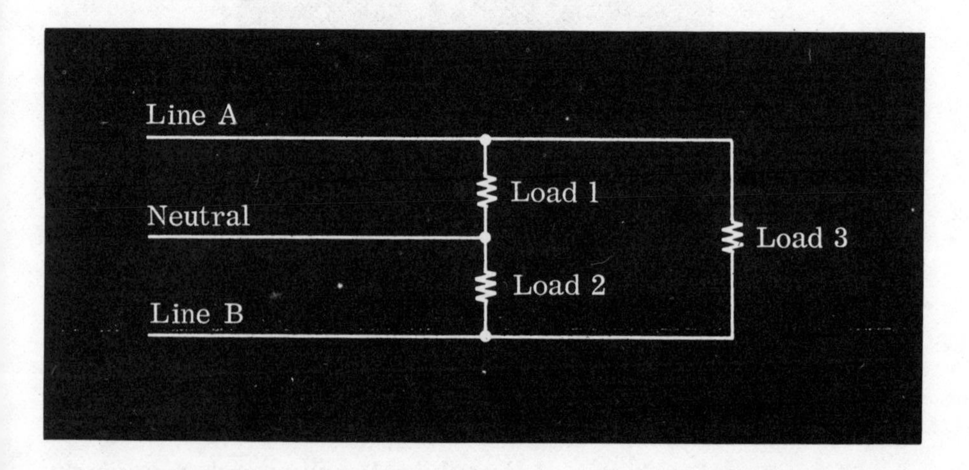

Remember that line A and line B are 180 degrees out of phase!

Troubleshooting Local Power Distribution Systems

Because of the way the circuits are connected to the lines, and because there are protective devices (fuses or circuit breakers about which you will learn more in Volume 5), it is relatively easy to isolate a circuit that has a fault. A blown fuse or open circuit breaker indicates an overload on a circuit. *You should never increase the size of the fuse or circuit breaker beyond the rated value.* Instead, if the line is simply overloaded from too many devices, you can move some of the loads to other circuits. If the overload persists, then either one of the devices on the circuit is defective or the line is shorted somewhere. This can be checked by removing the devices from the line and checking to see whether the overload still exists. A clamp-on ammeter is very useful in troubleshooting to allow for current measurement without disconnecting circuits.

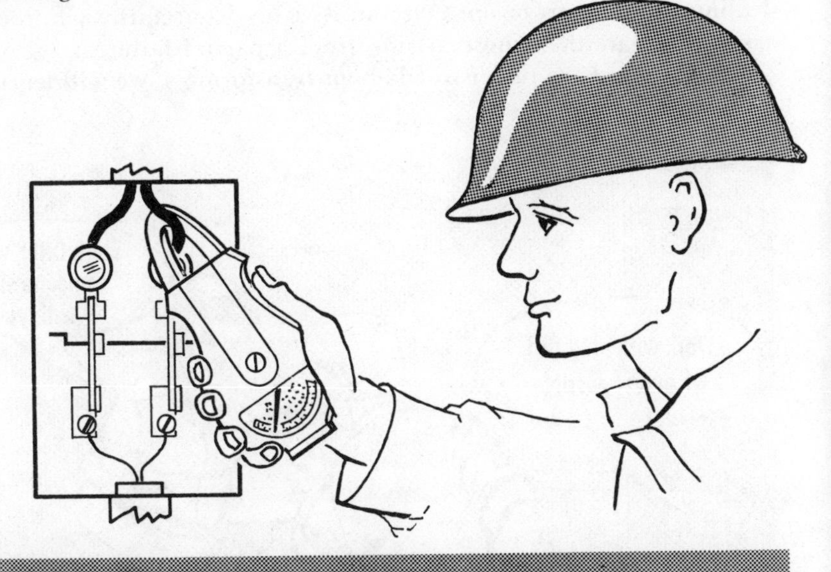

An open circuit can also easily be found because the voltage will be zero at all outlets or other convenient test points. If you know the way the wiring is arranged, you can trace the circuit to the point where the open circuit is and, more than likely, visually find the defect.

The importance of keeping the neutral line intact cannot be overemphasized! You could see this if you were to take the diagram on the previous page and put a very large load on one (low resistance) leg of the circuit and a very small load (high resistance) on the other leg. As long as the neutral is intact, the voltage across each load will be 120 volts and neutral current will flow. If you were to open the neutral, however, you would find that most of the voltage from the 240-volt line would appear across the smaller load, which, obviously, would burn out this device.

Introduction

You have already learned how to troubleshoot dc circuits in Volume 2 and also how to troubleshoot simple ac circuit elements (inductance and capacitance) in Volume 3. In this section, you will learn how to troubleshoot the ac circuits you have studied in this volume. As you learned in Volume 2, the important thing in troubleshooting is to use your head—that is, adopt a logical procedure and apply what you know. While hot components are not necessarily defective, the appearance of excessive heating can be a sure sign of difficulty. On the other hand, if components that should be at least warm are actually cold, then this condition may be equally suspicious. At this point, it is strongly recommended that you review the troubleshooting procedures described in Volumes 2 and 3. Use the ideas developed in Volume 2 to determine whether you are possibly dealing with a short or open circuit. As with dc circuits, such troubles are easier to locate than those arising from a partial failure.

Since you have just learned about transformers, we will learn how to troubleshoot them first.

Troubleshooting Transformer Circuits—Open Circuits

Since transformers are important parts of ac electrical equipment, you must know how to test them and how to locate faults that develop in them.

The three things that can cause transformer failures are (1) open circuit windings, (2) shorted windings, and (3) shorts to ground.

When one of the windings in a transformer develops an open circuit, no output current can flow and the transformer will not deliver any output. The symptom of an open-circuited transformer is, therefore, that the circuits which derive power from the transformer go dead. A check with an ac voltmeter across the transformer output terminals will show a reading of zero volts, although a voltmeter check across the transformer input terminals shows that voltage is present. Before you go further, this measurement is essential to make sure that the problem is not in the circuits feeding the transformer. Since there is voltage at the input and no voltage at the output, one of the windings must be open circuited. To find the fault, disconnect the transformer and check the windings for *continuity*, using an ohmmeter. Continuity (a continuous circuit) is indicated by a fairly low resistance reading, whereas an open winding will indicate an infinite resistance. In the majority of cases, the transformer will have to be replaced, unless of course the break is accessible and can be repaired.

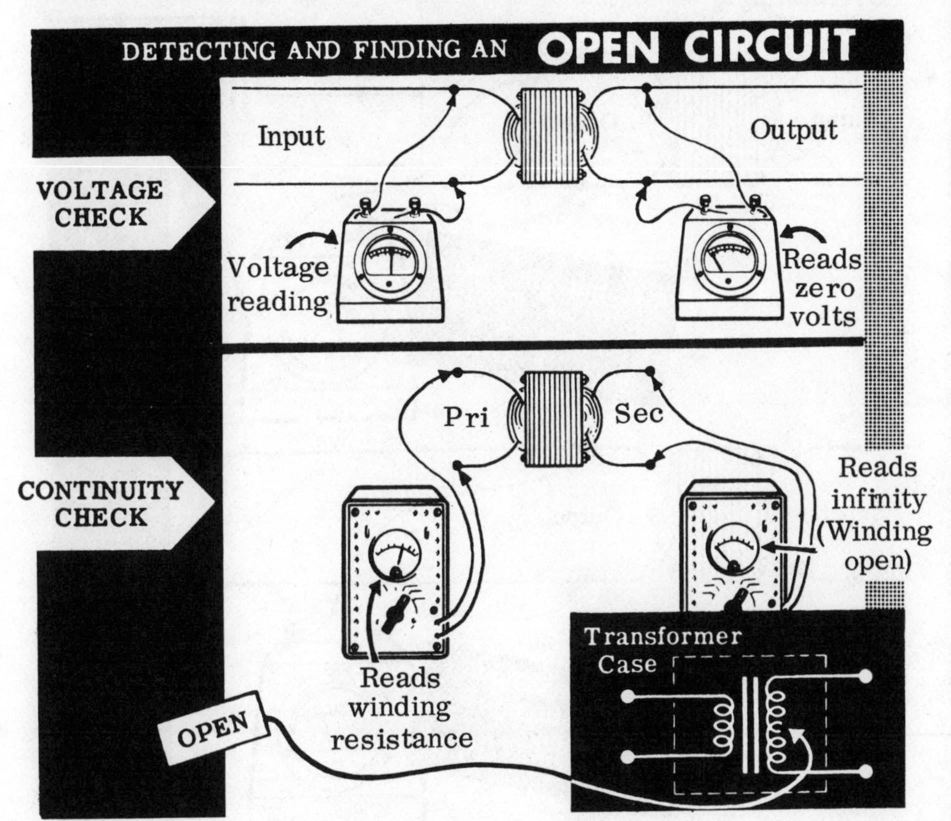

Troubleshooting Transformer Circuits—Short Circuits

When a few turns of a secondary winding of a transformer are shorted, the output voltage drops, and the transformer may overheat because of the increase in current in the secondary. The winding with the short may give a lower reading than normal on the ohmmeter. Usually the normal resistance reading is so low that a few shorted turns cannot be detected by using an ordinary ohmmeter. It is important to make sure that the problem is not in the load, which could be the reason why the transformer is drawing excessive current. One way to check this is to disconnect the load. If the transformer still overheats or the open circuit voltage is low, then it may be that the transformer has a shorted winding. In this case, a sure way to tell if the transformer is bad is to replace it with another equivalent transformer. If the replacement transformer operates satisfactorily, it should be used; the original transformer should be either repaired or discarded, depending on its size and type.

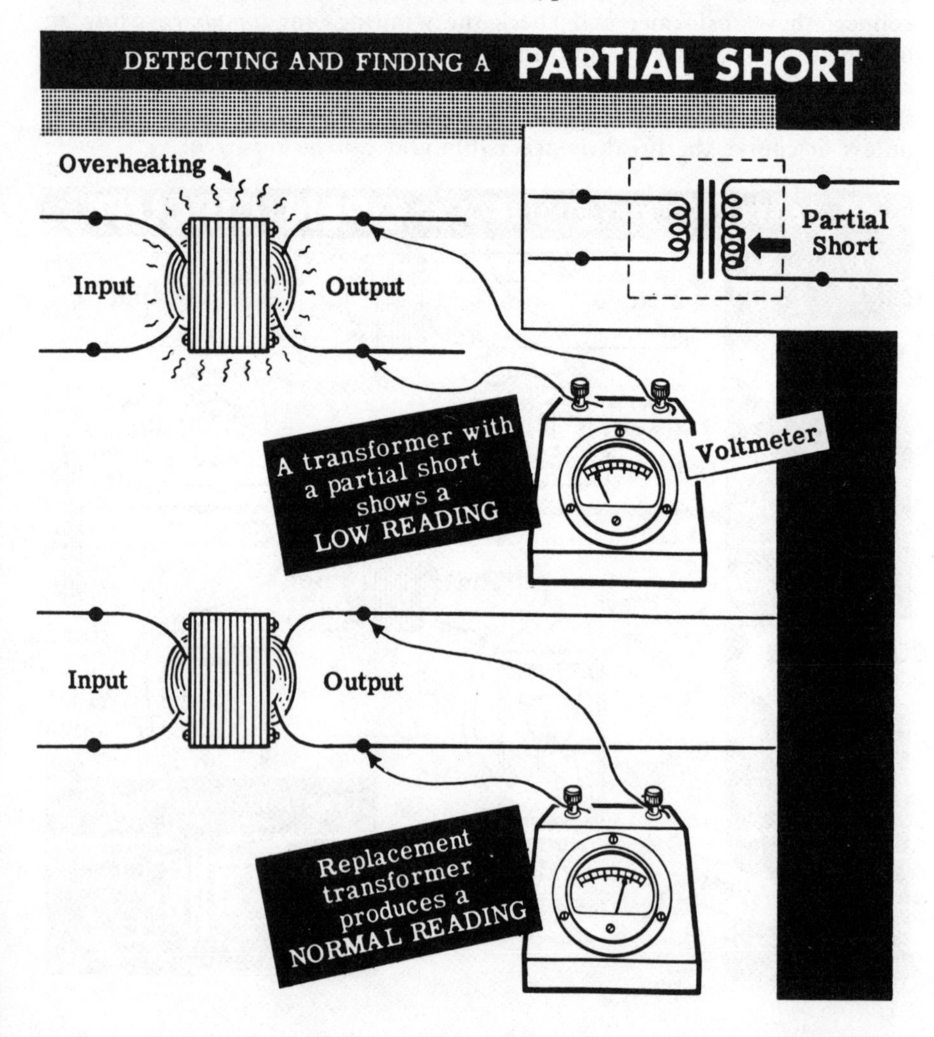

Troubleshooting Transformer Circuits—Short Circuits (continued)

Sometimes a winding has a complete short across it. The short may be in the external circuit connected to the winding or in the winding itself. Again, one of the symptoms is excessive overheating of the transformer because of the very large circulating current. The heat often melts the insulating materials inside the transformer, a fact which you can quickly detect by the smell. Also, there will be no voltage output across the shorted winding, and the circuit across the winding will be dead. In equipment which is fused, the heavy current flow will blow the fuse before the transformer is damaged completely. But if the fuse does not blow, the shorted winding may burn out.

The way to isolate the short is to disconnect the external circuit from the winding. If the voltage is normal with the external circuit disconnected, the short is in the external circuit. But if the voltage across the winding is still zero, it means the short is in the transformer, which will have to be repaired or replaced.

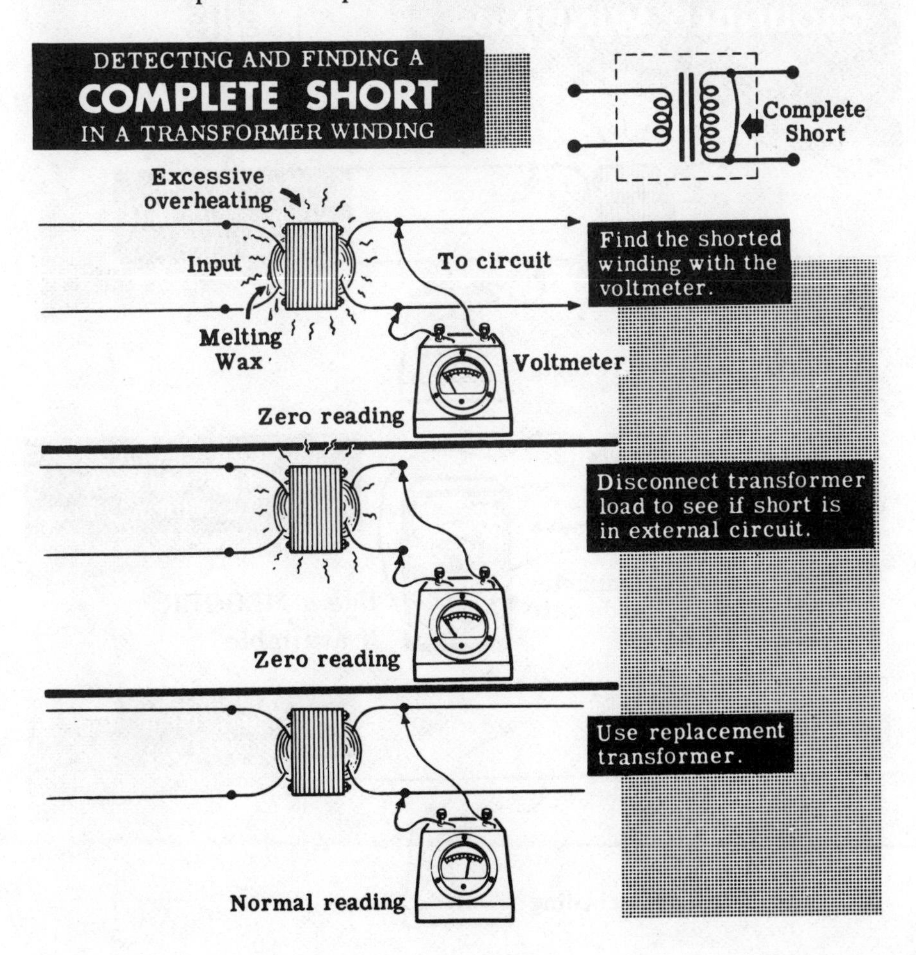

Troubleshooting Transformer Circuits—Short Circuits (continued)

Sometimes the insulation at some point in the winding breaks down, and the wire becomes exposed. It may touch the inside of the transformer case, shorting the wire to the case and thus grounding the winding.

If a winding develops a short to ground, and a point in the external circuit connected to this winding is also grounded, part of the winding will be shorted out. The symptoms will be the same as those for a shorted winding, and the transformer will have to be replaced or repaired. You can check whether a transformer winding is shorted to ground by connecting an ohmmeter between one side of the winding in question and the transformer case, but only after all the transformer leads have been disconnected from the circuit. A zero or low reading on the ohmmeter indicates that the winding is grounded.

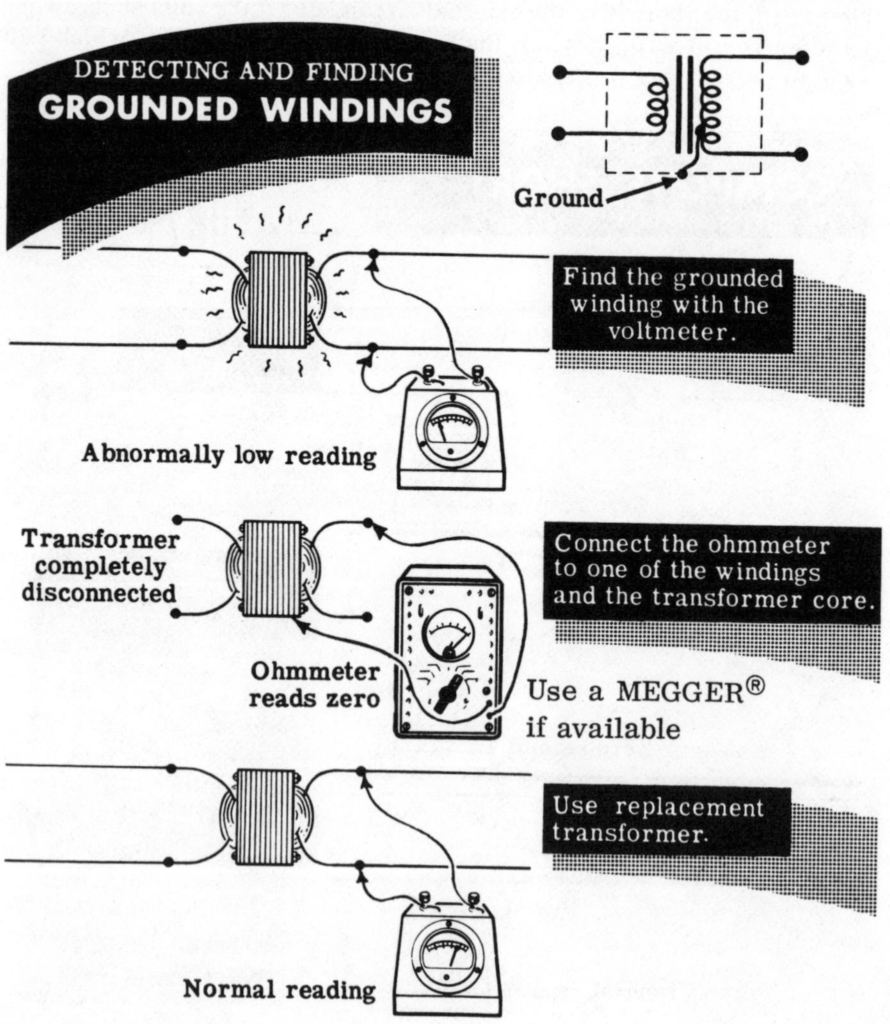

Troubleshooting AC Circuits

As for any troubleshooting, make sure that the input voltage is present and correct. In ac circuit work, you can use a clamp-on ammeter (see Volume 3) to measure circuit current without opening the circuit. You will find it a very convenient test instrument to have available. In addition, most of these meters have an auxiliary ac voltmeter that is very useful for checking circuit voltage.

Except for control applications, which you will learn about in Volume 5, series circuit connections are rarely used in ac circuits. In addition, you will find that complex ac circuits are found mainly in electronic systems. Usually, ac circuits are parallel connected. As you know, in parallel circuit connections the circuit voltage is common to all components (line voltage) but the currents depend on the component's impedance.

Suppose you had a line that was fused for 15 amperes (120 volts ac, 60 Hz) and, so far as you could tell, the normal load consisted of eight 100-watt lamps plus a small appliance that presented a load approximating that of a 50-ohm resistor in parallel with a 0.1-henry inductor.

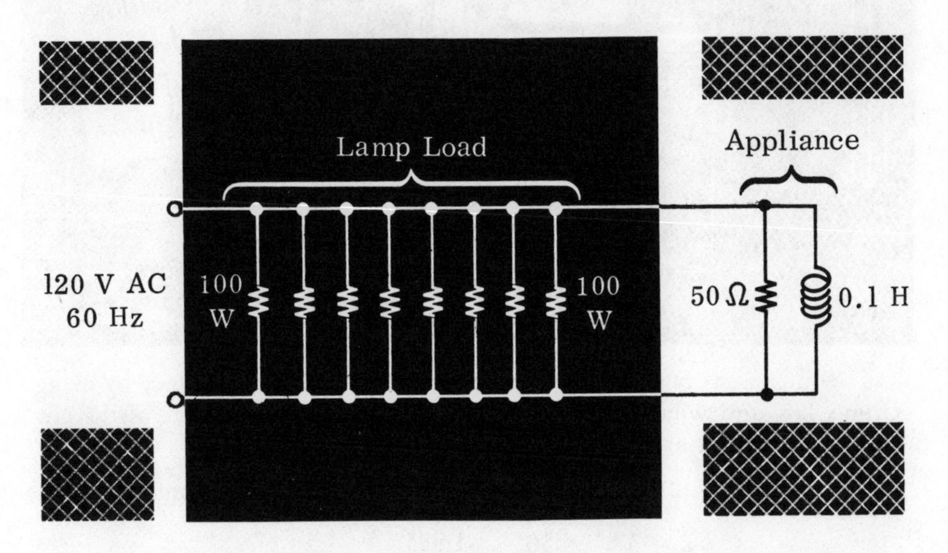

The fuse keeps blowing, indicating a circuit fault. With all the lamps and the appliance disconnected, the circuit is all right.

As you might suspect, a good way to find out which element is causing the difficulty would be to plug in each component one at a time to isolate the one at fault. You would, of course, leave the elements that function in the circuit, since the defective element alone might not draw enough current to cause the fuse to blow. As you return each circuit element to the line, check the current drawn by using the clamp-on ammeter.

You can use your knowledge of parallel circuits to figure out how much current each circuit component should draw.

Troubleshooting AC Circuits (continued)

Current Drawn by 100-Watt Lamp (Resistive)

$$P = EI$$

$$I = \frac{P}{E} = \frac{100}{120} = 0.8 \text{ A/lamp}$$

For 8 lamps, the load is $0.8 \times 8 = 6.4$ A

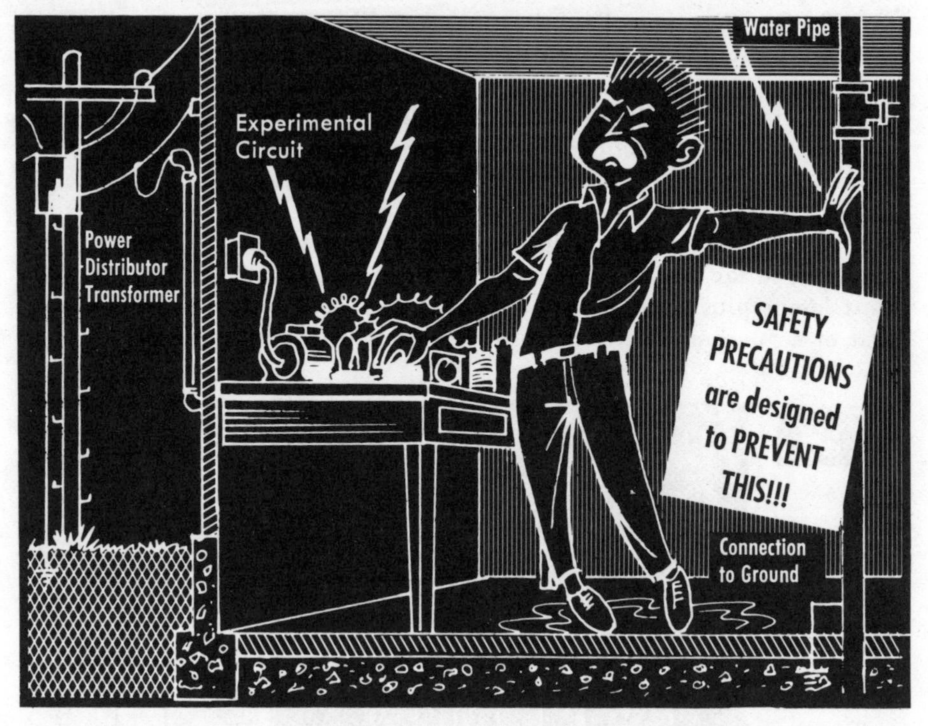

You can easily calculate the current drawn by the appliance by using Ohm's law and what you know about vectors. The impedance of the circuit is calculated as follows:

$$X_L = 2\pi fL = 2 \times \pi \times 60 \times 0.1 = 37.7 \text{ ohms}$$

$$Z = \frac{37.7 \times 50}{\sqrt{50^2 + 37.7^2}} = \frac{1{,}885}{62.6} = 30.11 \text{ ohms}$$

The current drawn by the appliance circuit is normally:

$$I = \frac{120}{30.11} = 3.99 \text{ A} \cong 4 \text{ A}$$

This, in combination with the lamp load, gives a total reading of

$$6.4 + 4 = 10.4 \text{ A}$$

Troubleshooting AC Circuits (continued)

A check with a clamp-on ammeter, or a series ammeter, shows that the appliance actually draws 10 amperes, which explains why the fuse blows.

The appliance can be checked out individually with an ohmmeter. In this case, the ohmmeter showed a partial short so that the inductor had a very low resistance. The problem was corrected by replacement of this part.

Remember that the fuse has to carry the *total* current and will blow at over 15 amperes, regardless of the phase angle. Thus, it is *volt-amperes* that need to be considered when a fuse or circuit breaker is involved. (However, only the in-phase component is involved with the real power used by the appliance.)

In later volumes, you will learn how to service the various electrical devices that are discussed. This special information coupled with what you know about electric circuits, and how to calculate them, will allow you to troubleshoot any electric circuit. More important than any particular item of test equipment, however, is *learning how to apply what you know.* Use your head and you will find that troubleshooting is not too difficult, even if you have just a few rudimentary tools. An essential part of troubleshooting is to know your electricity and, just as important, to know how the device or system you are troubleshooting works. The combination of this knowledge and skill will be the secret of your success in troubleshooting.

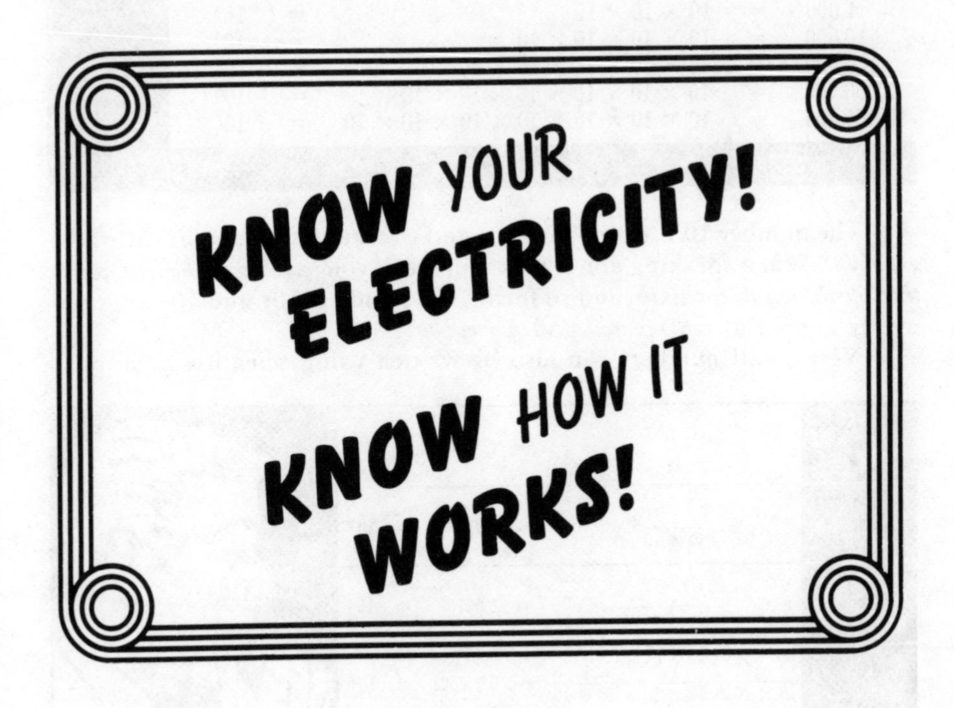

Scientific Notation

Your study so far has shown you that in working with electricity you will often use quantities as small as thousandths or millionths; in future work you may meet quantities as large as hundreds of thousands and even millions. While you have learned the rules for moving decimal points around to convert from one unit of measurement to another, there is a much simpler method available. That method is based upon understanding and using *scientific notation.*

Scientific notation is a shorthand method of writing down very large and very small numbers. Its use greatly simplifies making calculations with those numbers. This shorthand method makes use of the number 10 as the basis for writing down all large and small numbers.

The number of tens that have to be multiplied together to equal the desired number is written as a small numeral above and to the right of 10:

Scientific Notation

Desired Number		Number of Tens		Scientific Notation
1	=	Zero	=	10^0
10	=	10	=	10^1
100	=	10 × 10	=	10^2
1,000	=	10 × 10 × 10	=	10^3
10,000	=	10 × 10 × 10 × 10	=	10^4
100,000	=	10 × 10 × 10 × 10 × 10	=	10^5
1,000,000	=	10 × 10 × 10 × 10 × 10 × 10	=	10^6
10,000,000	=	10 × 10 × 10 × 10 × 10 × 10 × 10	=	10^7

The number 10 is called the *base*, and the smaller number is called the *exponent*. When speaking about these numbers, you say, *ten to the first, ten to the fourth, ten to the sixth*, and so forth. The numbers 10^2 and 10^3 are commonly known as *ten squared* and *ten cubed.*

Very small numbers can also be written using scientific notation:

$\frac{1}{10}$	=	$\frac{1}{10}$	=	0.1	=	10^{-1}
$\frac{1}{100}$	=	$\frac{1}{10 \times 10}$	=	0.01	=	10^{-2}
$\frac{1}{1000}$	=	$\frac{1}{10 \times 10 \times 10}$	=	0.001	=	10^{-3}
$\frac{1}{1{,}000{,}000}$	=	$\frac{1}{10 \times 10 \times 10 \times 10 \times 10 \times 10}$	=	0.000001	=	10^{-6}

Scientific Notation (continued)

You can use scientific notation to write down numbers smaller than 1 (one) by making the (negative) exponent equal to the count of zeros below the fraction line. For decimal numbers, the exponent is equal to one more than the number of places that the decimal point is located to the left of the first digit higher than zero. When speaking about these numbers you say *ten to the minus one, ten to the minus two, ten to the minus three,* and so forth. (You do not use the term *square* and *cube* when speaking of numbers smaller than one.)

Simply by following the concepts already explained, scientific notation can be used to indicate any number at all. For example:

The number to the left of the multiplication sign is the *multiplier.*

Although the examples that have been given should make the procedure for writing in scientific notation quite clear, there is a standard procedure that can be followed to resolve any questionable cases:

1. Write the desired quantity in terms of a decimal number.
2. Move the decimal point to the right or left depending on whether you are dealing with very large or very small numbers.
3. Count the number of places the decimal point was moved. This number is the exponent of the base 10.
4. If the decimal point was moved to the left, the exponent is positive (but no plus sign is required). If the decimal point was moved to the right, place a minus sign in front of the exponent.

Scientific Notation (continued)

In addition to simplifying the writing of large and small numbers, scientific notation makes it very easy to multiply and divide them. The rules are as follows:

Multiplication

1. *Multiply* the two *multipliers.* This gives you the new multiplier.
2. *Add* the two *exponents.* This gives you the new exponent.
3. Write the results in terms of scientific notation.
4. If necessary, move the decimal point to the right of the first digit larger than zero and adjust the value of the exponent. Refer to rules 3 and 4 on the previous page.

Example of Multiplication

Multiply 4×10^3 by 5.5×10^6
1. $4 \times 5.5 = 22.0$ (multiply)
2. $10^3 \times 10^6 = 10^{(3+6)} = 10^9$ (add exponents)
3. 22.0×10^9 (answer)
4. 2.2×10^{10} (answer adjusted)

Division

1. *Divide* the *first* multiplier (numerator) by the *second* (denominator). This gives you the new multiplier.
2. *Subtract* the denominator exponent from the numerator exponent. This gives you the new exponent.
3. Write the results in terms of scientific notation.
4. If necessary, adjust the value of the exponent. Refer to rules 3 and 4 on adjusting numbers.

Example of Division

Divide 3.6×10^9 by 6.0×10^3
1. $3.6 \div 6.0 = 0.6$
2. $10^9 \div 10^3 = 10^{(9-3)} = 10^6$
3. 0.6×10^6 (answer)
4. 6.0×10^5 (answer adjusted)

Note that $1/10^6$ is the same as 10^{-6}. You can prove this as follows:

$$\frac{1}{10^6} \times \frac{10^{-6}}{10^{-6}} = \frac{10^{-6}}{10^0} = \frac{10^{-6}}{1} = 10^{-6}$$

Scientific Notation (continued)

Conversely, $1/10^{-6} = 10^{6}$. Therefore, you can move factors-of-10 quantities from numerator to denominator or vice versa just by changing the sign of the exponent. You *cannot* do this with the multiplier; other rules of arithmetic hold in these cases.

You can also add *and* subtract numbers in scientific notation. However, *the value of the exponent must be the same before either operation:*

$3 \times 10^{4} + 4.7 \times 10^{4} = 7.7 \times 10^{4}$
$8.2 \times 10^{-5} + 2 \times 10^{-5} = 10.2 \times 10^{-5} = 1.02 \times 10^{-4}$

If you want to add or subtract numbers with different exponents, you must make the exponents *the same* before addition (or subtraction). For example, addition would be performed as follows:

$5.1 \times 10^{3} + 3.2 \times 10^{4} = 5.1 \times 10^{3} + 32 \times 10^{3} = 37.1 \times 10^{3}$
$= 3.71 \times 10^{-4}$

$3.5 \times 10^{-4} + 2.2 \times 10^{-3} = 3.5 \times 10^{-4} + 22 \times 10^{-4} = 25.5 \times 10^{-4} =$
2.55×10^{-3}

or you could do it as:

$3.5 \times 10^{-4} + 2.2 \times 10^{-3} = 0.35 \times 10^{-3} + 2.2 \times 10^{-3} = 2.55 \times 10^{-3}$

or, using subtraction:

$6.3 \times 10^{3} - 2.1 \times 10^{2} = 6.3 \times 10^{3} - 0.21 \times 10^{3} = 6.09 \times 10^{3}$
$5.1 \times 10^{3} - 3.2 \times 10^{4} = 5.1 \times 10^{3} - 32 \times 10^{3} = -26.9 \times 10^{3}$
$= -2.69 \times 10^{4}$

To find the square of a number in scientific notation, multiply the multiplier by itself and double the exponent.

Example

$(5 \times 10^{3})^{2} = 25 \times 10^{6}$
$(6.3 \times 10^{6}) = 39.69 \times 10^{12}$

To take the square root of a number in scientific notation, adjust the exponent so that it is an even number. Take the square root of the multiplier and halve the exponent.

Example

$\sqrt{5 \times 10^{3}} = \sqrt{50 \times 10^{2}} = 7.07 \times 10^{1} = 70.7$
$\sqrt{6.3 \times 10^{6}} = 2.51 \times 10^{3}$

Measurement Conversion

Earlier you learned how to convert units of current measurement by moving the decimal point. These procedures and your knowledge of scientific notation will make it a simple matter to convert other units.

By learning the simple mathematical relationships that exist between all units of electrical and electronic measures, you will not only be able to convert current measurements, but you will also be able to convert the measurements of volts, ohms, watts, farads, and henrys.

Measurement Conversion (continued)

All electrical, electronic, and many other types of scientific measurement make use of standard prefixes which are attached to the front of the word that is used as the standard unit of measure. These prefixes indicate the precise multiplier or fraction of that standard unit. You already know that the prefix "milli" means one-thousandth and the prefix "micro" means one-millionth. The range of prefixes in common use is as follows:

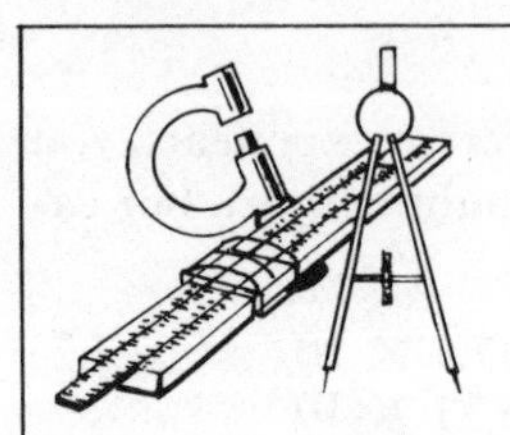

MEASUREMENT CONVERSION

Prefix	Abbrev.	Meaning	Mathematical Equivalent
pico (micromicro)	p ($\mu\mu$)	1 millionth of 1 millionth part of	10^{-12}
nano (millimicro)	n (mμ)	1 thousandth of 1 millionth part of	10^{-9}
micro	μ	1 millionth part of	10^{-6}
milli	m	1 thousandth part of	10^{-3}
centi	c	1 hundredth part of	10^{-2}
—	Unit	unit standard of measurement	10^{0}
kilo	k	1 thousand times	10^{3}
mega	M	1 million times	10^{6}
giga	G	1 thousand million times	10^{9}

For example, a thousandth of a volt is known as a *millivolt* and a million watts is known as a *megawatt*.

Measurement Conversion: Units of Current

Scientific notation makes it a simple matter to convert from one unit of measure to another, since the exponent tells you how the decimal point should be moved:

1. To change from any unit of measure to a *larger* one, move the decimal point to the *left* by a number of places equal to the difference between their exponents (as shown on the preceding page).
2. To change from any unit of measure to a *smaller* one, move the decimal point to the *right* by a number of places equal to the difference between their exponents (as shown on the preceding page).

To keep the explanation on familiar grounds, consider changing units of current measurement in the following examples.

Measurement Conversion: Units of Current (continued)

Example

Convert 3.5 milliampere (3.5 mA) to amperes.

Solution:
1. Milli = 10^{-3}; ampere (unit of measure) = 10^0
2. According to the rules, move the decimal point three places to the left.
3. 3.5 milliamperes = 0.003.5 ampere = 0.0035 ampere

Example

Convert 0.001 ampere to microamperes.

Solution:
1. Micro = 10^{-6}; ampere (unit of measure) = 10^0
2. According to the rules, move the decimal point six places to the right.
3. 0.001 ampere = .001000. microamperes = 1,000 μA

Example

Convert 2 milliamperes to microamperes.

Solution:
1. Micro = 10^{-6}; milli = 10^{-3}
2. According to the rules, move the decimal point three places to the right.
3. 2 milliamperes = 2.000. microamperes = 2,000 μA.

Measurement Conversion: Units of Voltage and Power

Units of voltage measurement are converted in exactly the same manner as units of current.

The exponents for the common units of voltage are as follows:

1 microvolt (μV) = 10^{-6}
1 millivolt (mV) = 10^{-3}
1 volt (V) = 10^0
1 kilovolt (kV) = 10^3
1 megavolt (MV) = 10^6.

Example

Convert 6,600 volts to kilovolts.

Solution:
1. Kilo = 10^3; volt (unit of measure) = 10^0
2. According to the rules, move the decimal point three places to the left.
3. 6,600 volts = 6.600. = 6.6 kilovolts

Measurement Conversion: Units of Voltage and Power (continued)

Likewise, similar conversions can be made with power units. The exponents for common units of power are:

1 microwatt (μW) = 10^{-6}
1 milliwatt (mW) = 10^{-3}
1 watt (W) = 10^{0}
1 kilowatt (kW) = 10^{3}
1 megawatt (MW) = 10^{6}.

Example

Convert 2.75 kilowatts to watts.

Solution:
1. Kilo = 10^{3}; watt (unit of measure) = 10^{0}
2. According to the rules, move the decimal point three places to the right.
3. 2.75 kilowatts = 2.750. = 2,750 watts

Measurement Conversion: Units of Resistance

Resistance is normally expressed in ohms. However, in some special applications, the resistance used may be as small as a fraction of an ohm or as large as many millions of ohms. This means that you may find resistance values expressed in microhms, milliohms, ohms, kilohms, and megohms and that you may have to convert from one unit of resistance measurement to another.

Units of resistance are converted in the same manner as units of current or voltage, by moving the decimal point as indicated by the exponent in scientific notation, and according to the table and rules of measurement conversion.

The exponents for the common units of resistance measurements are:

1 microhm ($\mu\Omega$) = 10^{-6}
1 milliohm (mΩ) = 10^{-3}
1 ohm (Ω) = 10^{0}
1 kilohm (K or KΩ) = 10^{3}
1 megohm (M or MΩ) = 10^{6}

Example

Convert 47 K to ohms.

Solution:
1. Kilo = 10^{3}; ohm (unit of measure) = 10^{0}
2. According to the rules, move the decimal point three places to the right.
3. 47 K = 47.000. = 47,000 ohms

Measurement Conversion: Units of Resistance (continued)

Example

Convert 1.2 megohms to kilohms.

Solution:
1. Mega = 10^6; kilohm = 10^3
2. According to the rules, move the decimal point three places to the right.
3. 1.2 megohms = 1.200. kilohms = 1,200 K

Applying Scientific Notation

Example A

Calculate the impedance of the following circuit at 1 MHz.

0.1 mH 1,000 pF 100 Ω

$$X_L = 2\pi fL = 2 \times 3.14 \times 10^6 \times 0.1 \times 10^{-3} = 628 \text{ ohms}$$

$$X_C = 1/2\pi fC = 1/2 \times \pi \times 10^6 \times 10^{-9} = 159 \text{ ohms}$$

$$Z = \sqrt{(100)^2 + (628 - 159)^2} = \sqrt{10^4 + 2.76 \times 10^5}$$

$$= \sqrt{10^4 + 27.6 \times 10^4} = \sqrt{28.6 \times 10^4} = 5.35 \times 10^2 = 535 \text{ ohms}$$

Example B

Calculate the resonant frequency of the parallel circuit below:

0.01 μF 0.05 H (50 mH)

$$f_r = \frac{1}{2\pi\sqrt{LC}}$$

$$= \frac{1}{2\pi\sqrt{50 \times 10^{-3} \times 0.01 \times 10^{-6}}}$$

$$= \frac{1}{2\pi\sqrt{50 \times 10^{-3} \times 10^{-8}}} = \frac{1}{2\pi\sqrt{50 \times 10^{-11}}}$$

$$= \frac{1}{6.28\sqrt{5 \times 10^{-10}}} = \frac{1}{6.28 \times 2.24 \times 10^{-5}}$$

$$= \frac{1}{14.067 \times 10^{-5}} = \frac{10^5}{14.067} = 7{,}108 \text{ Hz} = 7.108 \text{ kHz}$$

Example C

What are the apparent and true powers for a 235-kV transmission line carrying 200 amperes at a phase angle of 15 degrees?

Apparent power $= E \times I = 235 \text{ kV} \times 200 = 235 \times 10^3 \times 2 \times 10^2$
$= 470 \times 10^5$ watts = 47 MW

True power = apparent power $\times \cos\theta$
$= 47 \text{ MW} \times \cos\theta = 47 \text{ MW} \times 0.966 = 45.4 \text{ MW}$

Table of Squares and Square Roots

To use this table for values of N greater than 100, move the decimal point to the left two places at a time until a number less than 100 is obtained. Look up the square or square root as required for this number, then adjust your answer as follows: If the number is being squared, add *two* zeros to the left of the decimal point for each place you moved the decimal point originally. If the square root is being taken, move the decimal point one place to the right for each two places you originally moved it to the left.

Examples: What is the square of 900?

$900^2 = 9^2$ (moved decimal two places to the left) = 81 (move decimal *back* two places to the right for *each place* moved to the left originally) = 81 00 00 = 810,000.

What is the square root of 900?

$\sqrt{900} = \sqrt{9}$ (moved decimal two places to the left) = 3 (move decimal *back* one place to the right for *each two places* moved to the left originally) = 3 0 = 30.

N	N^2	$\sqrt{N}$	N	N^2	$\sqrt{N}$	N	N^2	$\sqrt{N}$
1	1	1.000	36	1,296	6.000	71	5,041	8.426
2	4	1.414	37	1,369	6.083	72	5,184	8.485
3	9	1.732	38	1,444	6.164	73	5,329	8.544
4	16	2.000	39	1,521	6.245	74	5,476	8.602
5	25	2.236	40	1,600	6.325	75	5,625	8.660
6	36	2.449	41	1,681	6.403	76	5,776	8.718
7	49	2.646	42	1,764	6.481	77	5,929	8.775
8	64	2.828	43	1,849	6.557	78	6,084	8.832
9	81	3.000	44	1,936	6.633	79	6,241	8.888
10	100	3.162	45	2,025	6.708	80	6,400	8.944
11	121	3.317	46	2,116	6.782	81	6,561	9.000
12	144	3.464	47	2,209	6.856	82	6,724	9.055
13	169	3.606	48	2,304	6.928	83	6,889	9.110
14	196	3.742	49	2,401	7.000	84	7,056	9.165
15	225	3.873	50	2,500	7.071	85	7,225	9.220
16	256	4.000	51	2,601	7.141	86	7,396	9.274
17	289	4.123	52	2,704	7.211	87	7,569	9.327
18	324	4.243	53	2,809	7.280	88	7,744	9.381
19	361	4.359	54	2,916	7.348	89	7,921	9.434
20	400	4.472	55	3,025	7.416	90	8,100	9.487
21	441	4.583	56	3,136	7.483	91	8,281	9.539
22	484	4.690	57	3,249	7.550	92	8,464	9.592
23	529	4.796	58	3,364	7.616	93	8,649	9.644
24	576	4.899	59	3,481	7.681	94	8,836	9.695
25	625	5.000	60	3,600	7.746	95	9,025	9.747
26	676	5.099	61	3,721	7.810	96	9,216	9.798
27	729	5.196	62	3,844	7.874	97	9,409	9.849
28	784	5.292	63	3,969	7.937	98	9,604	9.899
29	841	5.385	64	4,096	8.000	99	9,801	9.950
30	900	5.477	65	4,225	8.062	100	10,000	10.000
31	961	5.568	66	4,356	8.124			
32	1,024	5.657	67	4,489	8.185			
33	1,089	5.745	68	4,624	8.246			
34	1,156	5.831	69	4,761	8.307			
35	1,225	5.916	70	4,900	8.367			

Table of Sines (Sin), Cosines (Cos), and Tangents (Tan)

To find sin, cos, or tan, look up the nearest angle and read across to the right to the appropriate value. To find $\sin^{-1}$, $\cos^{-1}$, or $\tan^{-1}$, look for the closest value and read across to the left and find the angle. For more accuracy, you can estimate between values.

Angle	*Sin*	*Cos*	*Tan*
0	0	1	0
2.5	0.044	0.999	0.044
5	0.087	0.996	0.087
7.5	0.131	0.991	0.132
10	0.174	0.985	0.176
12.5	0.216	0.976	0.222
15	0.259	0.966	0.268
17.5	0.301	0.954	0.315
20	0.342	0.940	0.364
22.5	0.383	0.924	0.414
25	0.423	0.906	0.466
27.5	0.462	0.887	0.521
30	0.500	0.866	0.577
32.5	0.537	0.843	0.637
35	0.574	0.819	0.7
37.5	0.609	0.793	0.767
40	0.643	0.766	0.839
42.5	0.676	0.737	0.916
45	0.707	0.707	1.000
47.5	0.737	0.676	1.091
50	0.766	0.643	1.192
52.5	0.793	0.609	1.303
55	0.819	0.574	1.428
57.5	0.843	0.537	1.570
60	0.866	0.500	1.732
62.5	0.857	0.462	1.921
65	0.906	0.423	2.145
67.5	0.924	0.383	2.414
70	0.940	0.342	2.747
72.5	0.954	0.301	3.172
75	0.966	0.259	3.732
77.5	0.976	0.216	4.511
80	0.985	0.174	5.671
82.5	0.991	0.131	7.596
85	0.996	0.087	11.430
87.5	0.999	0.044	22.904
90	1.000	0	∞ (all reactance)

Learning Objectives—Next Volume

Overview—In the first four volumes of this course on basic electricity you have dealt with the fundamental principles of electricity and dc and ac electric circuits. To complete your study of electricity, you will learn how electrical machines make use of the properties of electricity and magnetism to generate electricity—generators—and to do mechanical work—motors. The construction and operation principles of generators, alternators, electric motors, and other dc and ac electrical devices and equipment will be described. Then you will see how these dc and ac generators and motors are connected to form a system and how various devices are used to control the system. In addition, you will learn how electric power is controlled and how it is used in its many applications. Also, you will learn something about solid-state devices that are used in electric power conversion, control, and distribution. Finally, you will be introduced to the maintenance and troubleshooting of dc and ac machinery and systems.

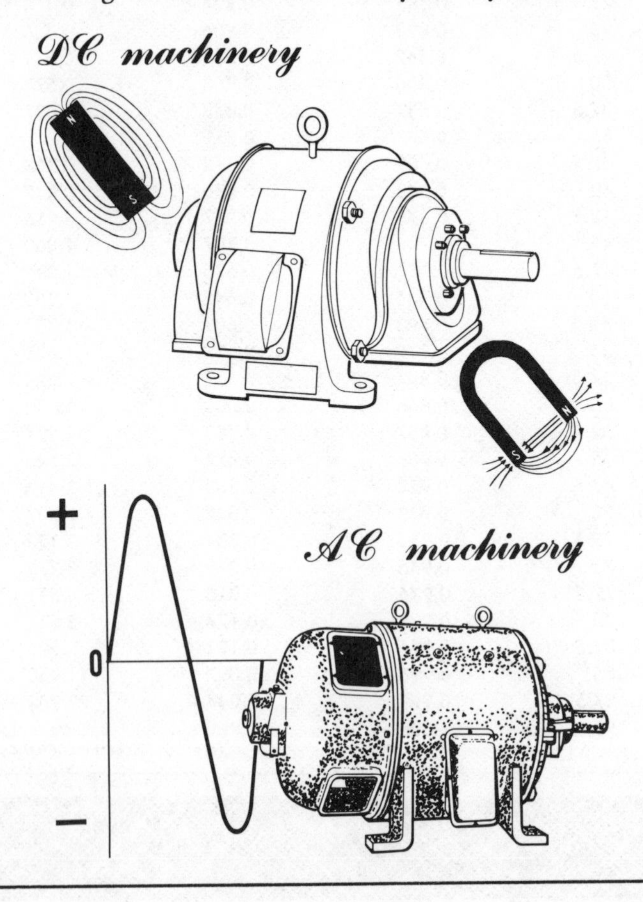

INDEX TO VOL. 4

(Note: A cumulative index covering all five volumes in this series will be found at the end of Volume 5.)